Integral Equations: A Short Course

Consulting Editor: Professor A. Jeffrey

Integral Equations: A Short Course

Ll. G. Chambers

INTERNATIONAL TEXTBOOK COMPANY LIMITED

Published by International Textbook Company Limited,
450 Edgware Road, London W2 1EG
A member of the Blackie Group

First published 1976

IBSN 0 7002 0262 5

Printed in Great Britain by Thomson Litho Ltd.,
East Kilbride, Scotland

Contents

Preface

This book arises out of a course of lectures I have given to honours mathematics students over the past few years. Although the book is, of necessity, aimed mainly at mathematicians, I have tried to keep excessive mathematical detail and abstractness out, and, in order not to hold up the development of the subject, a number of relevant mathematical topics have been relegated to appendices. A large number of worked examples and exercises are included, and it is hoped that these will be particularly useful to those studying by themselves. The background assumed is that of the usual second year mathematical methods course.

I have to thank Professor A. Jeffrey for a number of most useful comments on the first draft, Miss Ann Drybrough-Smith for her courtesy, and Mr. Owen Evans-Jones for reading the proofs. I am also grateful to the University of Wales for permission to use examination questions.

1 Preliminary Concepts

1.1 Introduction

The reader will no doubt be familiar with the idea of a differential equation. In this, some unknown function is defined in terms of a relation between its derivatives. From this relation it is possible, in theory, to obtain the unknown function which is defined apart from a number of arbitrary constants for ordinary differential equations, or a number of arbitrary functions for partial differential equations. The function is completely defined by specifying that some relations hold for it, or its derivatives, or a mixture of them at some points or over some domain. An integral equation is an equation in which the unknown function appears within an integral. It sometimes occurs that the same problem can be expressed both as an integral equation, and as a differential equation, but this is not always the case.

Consider therefore the simple problem defined by the differential equation

$$y'(x) = y(x), x \geqslant 0 \tag{1.1}$$

together with the condition

$$y(0) = 1 \tag{1.2}$$

The solution of this is clearly $\exp\{x\}$. Now integrate Eq. (1.1) with respect to x, and bring in the initial condition. The result is

$$y(x) - 1 = \int_0^x y(\xi)\mathrm{d}\xi \tag{1.3}$$

This is an integral equation into which the differential equation (1.1) and the initial condition have been transformed. It will be seen that the integral equation (1.3) becomes an identity on the substitution $y = \exp\{x\}$. Thus $\exp\{x\}$ is the solution of the integral equation (1.3) and the integral equation (1.3) may be regarded as an alternative formulation of the problem posed by the differential equation (1.1) and the initial condition (1.2).

Thus it is possible to formulate some problems both in differential equation form and integral equation form. Indeed, in some cases where numerical values are required, it may be advisable to transform a problem formulated in terms of a differential equation into an integral equation problem. The reason for this is linked with the fact that, in numerical analysis, differentiation increases error, but roughly speaking integration will tend to smooth errors out.

1.2 Some Problems Which Give Rise to Integral Equations

(a) *A Loaded Elastic String*

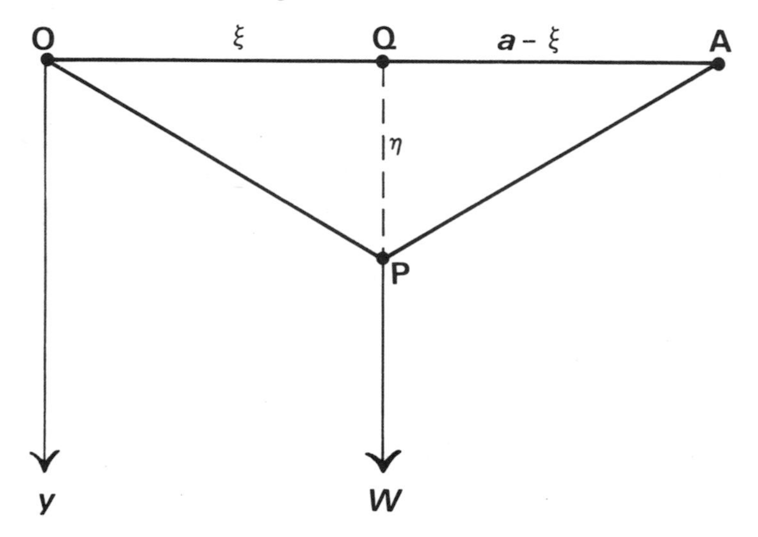

Consider a weightless elastic string (Fig. 1.1), stretched between two horizontal points O and A, and suppose that a weight W is hung from the elastic string and that in equilibrium the position of the weight is at a distance ξ from O and at a depth η below OA. If W is small, compared to the initial tension T in the string, it can be assumed that the tension of the string remains T during the further stretching. It is fairly clear that vertical resolution of forces gives the equilibrium equation

$$T(\eta/\xi)+T\{\eta/(a-\xi)\}-W=0 \tag{1.4}$$

where a is the length OA.

It follows that the drop η due to a weight W situated a distance ξ along the string from O is given by

$$\eta = W(a-\xi)\xi/(Ta) \tag{1.5}$$

It follows that the drop y in the string at a distance x from O is given by

$$y = x\eta/\xi \qquad 0 \leqslant x \leqslant \xi \tag{1.6a}$$

$$y = (a-x)\eta/(a-\xi) \qquad \xi \leqslant x \leqslant a \tag{1.6b}$$

These two results can be written in the form

$$y = WG(x, \xi)/T \tag{1.7}$$

where

$$G(x, \xi) = x(a-\xi)/a \qquad 0 \leqslant x \leqslant \xi \tag{1.8a}$$

$$= \xi(a-x)/a \qquad \xi \leqslant x \leqslant a \tag{1.8b}$$

Suppose now that the string is loaded continuously with a weight distribution $w(x)$ per unit length. Then the elementary displacement at the point distant x from 0, due to the weight distribution over $\xi \leqslant x \leqslant \xi+\delta\xi$, is

$$\delta y = w(\xi)\delta\xi G(x, \xi)/T \qquad 0 \leqslant x, \xi \leqslant a \tag{1.9}$$

It follows on integrating that the displacement due to the complete weight distribution is given by

$$y(x) = T^{-1}\int_0^a G(x, \xi)w(\xi)\mathrm{d}\xi \qquad 0 \leqslant x \leqslant a \tag{1.10}$$

In this way, the displacement of the string is given in terms of the weight distribution. However, the system might be looked at in another way. Given the displacement of the string, what is the weight distribution? The answer to this question is given by the solution of the integral equation (1.10) for the unknown function w. In this particular case, the solution of the integral equation is quite easy, as Eq. (1.10) may be rewritten in the form

$$y(x) = (Ta)^{-1}\left[x\int_0^x (a-\xi)w(\xi)\mathrm{d}\xi + (a-x)\int_x^a \xi w(\xi)\mathrm{d}\xi\right] \tag{1.11}$$

It may be verified, without any great difficulty, that a double differentiation of Eq. (1.11) with respect to x will give

$$y''(x) = (Ta)^{-1}w(x)$$

whence

$$w(x) = Tay''(x)$$

and this is the solution of the integral equation (1.10).

(b) The Shop Stocking Problem

A shop starts selling some goods. It is found that a proportion $K(t)$ remains unsold at time t after the shop has purchased the goods. It is required to find the rate at which the shop should purchase the goods, so that the stock of the goods in the shop remains constant (all processes are deemed to be continuous).

Suppose that the shop commences business in the goods by purchasing an amount A of the goods at zero time, and buys at a rate $\phi(t)$ subsequently. Over the time interval $\tau \leqslant t \leqslant \tau+\delta\tau$, an amount $\phi(\tau)\delta\tau$ is bought by the shop, and at time t the portion of this remaining unsold is

$$K(t-\tau)\phi(\tau)\delta\tau$$

Thus the amount of goods remaining unsold at time t, and which was bought up to that time, is given by

$$AK(t)+\int_0^t K(t-\tau)\phi(\tau)\mathrm{d}\tau$$

This is the total stock of the shop and is to remain constant at its initial value and so

$$A = AK(t)+\int_0^t K(t-\tau)\phi(\tau)\mathrm{d}\tau$$

and the required restocking rate $\phi(t)$ is the solution of this integral equation.

(c) A Problem in Electrostatics

The two problems discussed in Sections (a) and (b) have involved problems which arise in one dimension only. It is worth mentioning that, just as there are partial differential equations which involve differentiating with respect to more than one variable, there are integral equations which involve integrating over more than one variable. A fairly simple example of this arises in electrostatics. Without going into very great detail,[1] it can be shown that, if there is an electric charge distribution of density $\rho(r)$, then, provided that the charge distribution is sufficiently small at great distances from the origin, the electrostatic potential is given by

$$V(r) = (4\pi\varepsilon)^{-1}\int\frac{\rho(r')\mathrm{d}\tau'}{|r-r'|} \tag{1.12}$$

(ε is the permittivity)

However, Eq. (1.12) may be regarded as an integral equation to determine an unknown charge distribution $\rho(r)$ in terms of a known potential distribution $V(r)$. The solution is, in fact, well known, being given by

$$\rho(r) = -\varepsilon\nabla^2 V \tag{1.13}$$

Alternatively, the solution of the differential equation (1.13), with the additional condition that V should be sufficiently small at great distances, is given by Eq. (1.12).

(d) An Eigenvalue Problem

Consider the eigenvalue problem defined by the differential equation

$$\frac{d^2y}{dx^2} + \lambda y = 0 \qquad 0 \leqslant x \leqslant a \tag{1.14}$$

and the boundary conditions

$$y = 0 \quad \text{at} \quad x = 0 \quad \text{and} \quad x = a \tag{1.15}$$

Multiplying by y, and integrating over $0 \leqslant x \leqslant a$, it follows that

$$-\int_0^a y\frac{d^2y}{dx^2}dx = \lambda\int_0^a y^2 dx$$

and integrating by parts, it follows that

$$\int_0^a \left(\frac{dy}{dx}\right)^2 dx = \lambda\int_0^a y^2 dx \tag{1.16}$$

This is an example of an eigenvalue problem associated with an integral equation. The solution of the problem posed by the relations (1.14) and (1.15) is given by

$$y = \sin(n\pi x/a),\ \lambda = n^2\pi^2/a^2,\ n \text{ being a positive integer} \tag{1.17}$$

and it can be verified that Eq. (1.16) is satisfied.

If however there is the same differential equation, but different boundary conditions

$$\frac{dy}{dx} = 0 \quad \text{at} \quad x = 0 \quad \text{and} \quad x = a \tag{1.18}$$

the integral equation for y is again Eq. (1.16), but the solution is given by

$$y = \cos(n\pi x/a),\ \lambda = n^2\pi^2/a^2,\ n \text{ being zero or a positive integer} \tag{1.19}$$

Thus the boundary conditions are of importance in considering the solution of the problem posed by the integral equation (1.16). It may be observed, and the reader will easily verify it, that the problem posed by the relations (1.14) and (1.15) may be reformulated as

$$y(x) + \lambda a^2 \int_0^a G(x, \xi)y(\xi)d\xi = 0 \tag{1.20}$$

where $G(x, \xi)$ is defined by Eqs. (1.8).

1.3 Conversion of Ordinary Differential Equations into Integral Equations

(a) Linear Differential Equations with Initial Conditions (Marching Problems)

Consider the differential equation

$$y''(x)+a_1(x)y'(x)+a_2(x)y(x) = f(x) \tag{1.21}$$

with the initial conditions

$$y(0) = y_0,\ y'(0) = y_1 \tag{1.22}$$

Such a problem is, in numerical analysis, termed a marching problem because the solution so to speak marches out from the origin.

Let

$$\psi(x) = y''(x) \tag{1.23a}$$

Then

$$y'(x) = \int_0^x \psi(u)\mathrm{d}u + y_1 \tag{1.23b}$$

$$y(x) = \int_0^x (x-u)\psi(u)\mathrm{d}u + y_1 x + y_0 \tag{1.23c}$$

Substituting the relations (1.23) into the differential equation, it follows that

$$\psi(x)+\int_0^x [a_1(x)+a_2(x)(x-u)]\psi(u)\mathrm{d}u = f(x)-y_1a_1(x)-y_1xa_2(x)-y_0a_2(x) \tag{1.24}$$

Equation (1.24) can be written in the form

$$\psi(x)+\int_0^x K(x,u)\psi(u)\mathrm{d}u = g(x) \tag{1.25}$$

which is an integral equation for $\psi(x)$. y can be obtained by the use of the relations (1.22) and (1.23).

The ideas indicated here extend to a differential equation

$$\sum_{p=0}^{n} a_{n-p}(x)y^{(p)}(x) = f(x) \qquad (a_0 = 1) \tag{1.26}$$

by making use of the result

$$\left[\int_{x_0}^{x}\right]^n f(\xi)\mathrm{d}\xi = \int_{x_0}^{x} \frac{(x-\xi)^{n-1}}{(n-1)!} f(\xi)\mathrm{d}\xi \tag{1.27}$$

(see Appendix A)

Example 1.1

Form the integral equation corresponding to $y''+2xy'+y = 0$, $y(0) = 1$, $y'(0) = 0$.

Let

$$y'' = \psi$$

$$y' = \int_0^x \psi(u)\mathrm{d}u, \; y = \int_0^x (x-u)\psi(u)\mathrm{d}u+1$$

Thus

$$\psi+2x\int_0^x \psi(u)\mathrm{d}u+\int_0^x (x-u)\psi(u)\mathrm{d}u+1 = 0$$

whence

$$\psi(x)+\int_0^x (3x-u)\psi(u)\mathrm{d}u+1 = 0$$

(b) Transformation of Sturm-Liouville Problems to Integral Equations

A problem which is associated with an expression of the form

$$Ly = \frac{\mathrm{d}}{\mathrm{d}x}\left\{p(x)\frac{\mathrm{d}y}{\mathrm{d}x}\right\}-q(x)y \qquad x_1 \leqslant x \leqslant x_2 \tag{1.28}$$

and boundary conditions of the form

$$\begin{aligned} a_1y(x_1)+b_1y'(x_1) &= 0 \\ a_2y(x_2)+b_2y'(x_2) &= 0 \end{aligned} \tag{1.29}$$

is said to be of Sturm-Liouville type.

There are two problems which are of interest here, namely

$$Ly = f(x) \qquad x_1 \leqslant x \leqslant x_2 \tag{1.30}$$

and

$$Ly+\lambda r(x)y = 0 \qquad x_1 \leqslant x \leqslant x_2 \tag{1.31}$$

$p(x)$, $q(x)$ and $r(x)$ are continuous in the interval $x_1 \leqslant x \leqslant x_2$, and in addition $p(x)$ has a continuous derivative and does not vanish.

The differential equation (1.30) corresponds to a displacement y caused by some forcing function f, and the differential equation (1.31) forms, together with the boundary conditions, an eigenvalue problem.

Before proceeding with the transformation of the differential equations to integral form, consider a property of the differential equation

$$Ly = 0 \tag{1.32}$$

Suppose that ϕ_1, ϕ_2 are two independent solutions of this equation. Then

$$\begin{aligned} 0 &= \phi_2 L\phi_1 - \phi_1 L\phi_2 \\ &= \phi_2 \frac{\mathrm{d}}{\mathrm{d}x}\left(p\frac{\mathrm{d}\phi_1}{\mathrm{d}x}\right) - \phi_1 \frac{\mathrm{d}}{\mathrm{d}x}\left(p\frac{\mathrm{d}\phi_2}{\mathrm{d}x}\right) \\ &= \frac{\mathrm{d}}{\mathrm{d}x}\left\{p\left(\phi_2\frac{\mathrm{d}\phi_1}{\mathrm{d}x} - \phi_1\frac{\mathrm{d}\phi_2}{\mathrm{d}x}\right)\right\} \end{aligned}$$

and it follows that

$$p\left(\phi_2\frac{\mathrm{d}\phi_1}{\mathrm{d}x} - \phi_1\frac{\mathrm{d}\phi_2}{\mathrm{d}x}\right) = \text{const.} \tag{1.33}$$

This relation will be needed later.

Consider now the solution of the differential equation (1.30) subject to the boundary conditions (1.29). Suppose that $\phi_1(x)$ and $\phi_2(x)$ are two independent solutions of the differential equation $Ly = 0$ and that

$$\begin{aligned} a_1\phi_1(x_1) + b_1\phi_1'(x_1) &= 0 \\ a_2\phi_2(x_2) + b_2\phi_2'(x_2) &= 0 \end{aligned} \tag{1.34}$$

It is assumed that it is not possible to obtain a solution of $Ly = 0$ which satisfies both boundary conditions simultaneously. Such a solution would be a solution of the differential equation (1.31) corresponding to a zero eigenvalue. This case will be dealt with separately.

Using the method of variation of parameters, look for a solution of Eq. (1.30) of the form

$$y(x) = z_1(x)\phi_1(x) + z_2(x)\phi_2(x) \tag{1.35}$$

where z_1 and z_2 are to be determined. It follows that

$$y' = z_1'\phi_1 + z_2'\phi_2 + z_1\phi_1' + z_2\phi_2'$$

Let

$$z_1'\phi_1 + z_2'\phi_2 = 0 \tag{1.36}$$

Then

$$\begin{aligned} Ly &= \frac{\mathrm{d}}{\mathrm{d}x}[p(x)\{z_1(x)\phi_1'(x) + z_2(x)\phi_2'(x)\}] - q(x)\{z_1(x)\phi_1(x) + z_2(x)\phi_2(x)\} \\ &= p(z_1'\phi_1' + z_2'\phi_2') \end{aligned} \tag{1.37}$$

on using the results $L\phi_1 = L\phi_2 = 0$.

Thus z_1, z_2 are given by the solutions of equations

$$\begin{aligned} z_1'\phi_1 + z_2'\phi_2 &= 0 \\ p(z_1'\phi_1' + z_2'\phi_2') &= f \end{aligned}$$

which make the expression for y given by Eq. (1.35) satisfy the boundary conditions. Thus

$$z_1' = \frac{f\phi_2}{p(\phi_2\phi_1' - \phi_1\phi_2')}, \quad z_2' = -\frac{f\phi_1}{p(\phi_2\phi_1' - \phi_1\phi_2'}$$

The denominator in these two expressions is constant by the result (1.33) and by a suitable scaling of ϕ_1 and ϕ_2 may be taken as -1. Thus

$$z_1' = -f\phi_2, \; z_2' = f\phi_1 \tag{1.38}$$

It follows that

$$z_1(x) = \int_x^{\alpha} \phi_2(\xi) f(\xi) \mathrm{d}\xi$$

$$z_2(x) = \int_{\beta}^{x} \phi_1(\xi) f(\xi) \mathrm{d}\xi$$

where the unspecified limits of integration are the equivalent of the arbitrary constants of integration and are determined by the necessity of y satisfying the boundary conditions. Now

$$a_1 y + b_1 y' = a_1(z_1\phi_1 + z_2\phi_2) + b_1(z_1\phi_1' + z_2\phi_2')$$

using Eq. (1.36). Also

$$a_1\phi_1(x_1) + b_1\phi_1'(x_1) = 0$$

hence

$$0 = a_1 y(x_1) + b_1 y'(x_1) = z_2(x_1)[a_1\phi_2(x_1) + b_1\phi_2'(x_1)]$$

Now it has been assumed that neither ϕ_1 nor ϕ_2 satisfies both boundary conditions, hence it follows that $z_2(x_1) = 0$ and so

$$z_2(x) = \int_{x_1}^{x} \phi_1(\xi) f(\xi) \mathrm{d}\xi \tag{1.39a}$$

Similarly

$$z_1(x) = \int_x^{x_2} \phi_2(\xi) f(\xi) \mathrm{d}\xi \tag{1.39b}$$

Thus, it follows that

$$y(x) = \int_{x_1}^{x_2} G(x, \xi) f(\xi) \mathrm{d}\xi \tag{1.40}$$

where

$$\begin{aligned} G(x, \xi) &= \phi_1(\xi)\phi_2(x) \qquad x_1 \leqslant \xi \leqslant x \\ &= \phi_1(x)\phi_2(\xi) \qquad x \leqslant \xi \leqslant x_2 \end{aligned} \tag{1.41}$$

The quantity $G(x,\xi)$ is termed the Green's function associated with the operator L and the boundary conditions specified. It will be noted that G is symmetric in x and ξ. Equation (1.40) is thus an integral equation for f, the solution of which is given by the differential equation (1.32).

It follows immediately that the eigenvalue problem defined by the differential equation (1.31) and the boundary conditions (1.29) can be reformulated as the integral equation

$$y(x)+\lambda\int_{x_1}^{x_2} G(x,\xi)r(\xi)y(\xi)\mathrm{d}\xi = 0 \tag{1.42}$$

The above analysis has assumed that the solutions ϕ_1, ϕ_2 of $Ly = 0$ do not satisfy both boundary conditions. Suppose that ϕ is a solution which does in fact satisfy both boundary conditions, that is, it is an eigensolution corresponding to a zero eigenvalue in the differential equation (1.31). Now the differential equation $Ly = 0$ is of the second order and so has two independent solutions. Suppose that ψ is a second solution. This will not satisfy either boundary condition. Then, following the previous process, it follows that

$$y(x) = \phi(x)\int_{\alpha}^{x}\psi(\xi)f(\xi)\mathrm{d}\xi+\psi(x)\int_{x}^{\beta}\phi(\xi)f(\xi)\mathrm{d}\xi \tag{1.43}$$

where α and β are arbitrary. Following through the same processes as previously, and remembering that y and ϕ satisfy both boundary conditions, it follows that

$$0 = a_1y(x_1)+b_1y'(x_1) = [a_1\psi_1(x_1)+b_1\psi'_1(x_1)]\int_{x_1}^{\beta}\phi(\xi)f(\xi)\mathrm{d}\xi \tag{1.44}$$

$$0 = a_2y(x_2)+b_2y'(x_2) = [a_2\psi_2(x_2)+b_2\psi'_2(x_2)]\int_{x_2}^{\beta}\phi(\xi)f(\xi)\mathrm{d}\xi$$

Now ψ does not satisfy either boundary condition and so it follows from the first equation that $\beta = x_1$ and from the second one that

$$\int_{x_1}^{x_2}\phi(\xi)f(\xi)\mathrm{d}\xi = 0 \tag{1.45}$$

and a solution is only possible when there is this relation between f and ϕ. Thus the integral equation formulation becomes

$$y = A\phi(x)+\int_{x_1}^{x_2} G(x,\xi)f(\xi)\mathrm{d}\xi \tag{1.46}$$

where A is an arbitrary constant

$$= \int_{\alpha}^{x_1}\psi(\xi)f(\xi)\mathrm{d}\xi \tag{1.47}$$

and

$$G(x, \xi) = \phi(\xi)\psi(x) \qquad x_1 \leqslant \xi \leqslant x$$
$$= \phi(x)\psi(\xi) \qquad x \leqslant \xi \leqslant x_2 \tag{1.48}$$

There is thus ambiguity in the formulation as an integral equation in f.

Example 1.2

Find an integral equation formulation for the problem defined by

$$\frac{d^2y}{dx^2} + 4y = f(x) \qquad 0 \leqslant x \leqslant \pi/2$$

$$y = 0 \text{ at } x = 0, \text{ and } y = 0 \text{ at } x = \pi/2$$

It is easy to see that the solutions of

$$\frac{d^2y}{dx^2} + 4y = 0$$

which satisfy the boundary conditions at $x = 0$ and $x = \pi/2$ are respectively $\sin 2x$ and $\cos 2x$. Neither satisfies both boundary conditions.

Let $\quad y = w \sin 2x + z \cos 2x$

$$y' = w' \sin 2x + z' \cos 2x + 2w \cos 2x - 2z \sin 2x$$
$$= 2w \cos 2x - 2z \sin 2x$$

if $\quad w' \sin 2x + z' \cos 2x = 0$

$$y'' = 2w' \cos 2x - 2z' \sin 2x - 4w \sin 2x - 4z \cos 2x$$

Thus the equation $y'' + 4y = f$, becomes

$$2w' \cos 2x - 2z' \sin 2x = f$$

whence $\qquad z' = -\tfrac{1}{2} f \sin 2x, \; w' = \tfrac{1}{2} f \cos 2x$

Thus

$$2y = \sin 2x \int^x \cos 2\xi f(\xi) d\xi + \cos 2x \int_x \sin 2\xi f(\xi) d\xi$$

where the limits of integration are as yet unspecified

$$y' = \cos 2x \int^x \cos 2\xi f(\xi) d\xi - \sin 2x \int_x \sin 2\xi f(\xi) d\xi$$

y is to vanish at $x = 0$, so the unspecified limit of integration in the second integral is zero. Similarly, because y vanishes at $x = \pi/2$, the limit of integration to be specified in the first integral is $\pi/2$. Thus

$$y(x) = \int_0^{\pi/2} G(x, \xi) f(\xi) \mathrm{d}\xi$$

where

$$\begin{aligned} G(x, \xi) &= -\tfrac{1}{2}\cos 2\xi \sin 2x \qquad 0 \leqslant x \leqslant \xi \\ &= -\tfrac{1}{2}\cos 2x \sin 2\xi \qquad \xi \leqslant x \leqslant \frac{\pi}{2} \end{aligned}$$

Example 1.3

Transform the problem defined by

$$\frac{\mathrm{d}^2 y}{\mathrm{d}x^2} + \lambda y = 0$$

when $y = 0$ at $x = 0$, and $y' = 0$ at $x = 1$ into integral equation form.

The answer to the problem is in fact

$$y = \sin\frac{(2r-1)\pi x}{2} \qquad \lambda = \left[\frac{(2r-1)\pi}{2}\right]^2$$

r being a positive integer.

The two solutions of

$$\frac{\mathrm{d}^2 y}{\mathrm{d}x^2} = 0$$

which satisfy the boundary conditions are respectively $y = x$ and $y = 1$.

Following through the usual processes, it follows that the solution of

$$\frac{\mathrm{d}^2 y}{\mathrm{d}x^2} = f(x)$$

under the boundary conditions specified is

$$y = x\int_1^x f(\xi)\mathrm{d}\xi + \int_x^0 \xi f(\xi)\mathrm{d}\xi$$

and so the integral equation formulation is

$$y(x) = \lambda\int_0^1 K(x, \xi) y(\xi)\mathrm{d}\xi$$

where

$$\begin{aligned} K(x, \xi) &= x \qquad 0 \leqslant x \leqslant \xi \\ &= \xi \qquad \xi \leqslant x \leqslant 1 \end{aligned}$$

Example 1.4

Transform the problem defined by

$$\frac{d^2y}{dx^2}+y = f(x)$$

and the boundary conditions $y = 0$ at $x = 0$ and $x = \pi$ into integral equation form and indicate what condition must be satisfied by $f(x)$.

Clearly $\sin x$ satisfies the differential equation

$$\frac{d^2y}{dx^2}+y = 0$$

and both boundary conditions.

A second solution of the differential equation

$$\frac{d^2y}{dx^2}+y = 0$$

is $\cos x$, and this satisfies neither boundary condition.

Let

$$y = z\sin x + w\cos x$$

Following through the same processes as in Example 1.2, it follows that

$$y = \sin x \int^{x} \cos\xi f(\xi)\, d\xi + \cos x \int_{x} \sin\xi f(\xi)\, d\xi$$

Now y is to vanish at $x = 0$, and so the limit of integration on the second integral is zero.

y must also vanish at $x = \pi$ and it follows therefore that

$$y(\pi) = \cos\pi \int_{\pi}^{0} \sin\xi f(\xi)\, d\xi$$

Thus, for a solution to be possible,

$$\int_{0}^{\pi} \sin\xi f(\xi)\, d\xi = 0$$

and

$$y(x) = A\sin x + \int_{0}^{\pi} G(x, \xi) f(\xi)\, d\xi$$

where A is arbitrary and

$$\begin{aligned} G(x, \xi) &= -\sin\xi\cos x \qquad 0 \leqslant \xi \leqslant x \\ &= -\sin x\cos\xi \qquad x \leqslant \xi \leqslant \pi \end{aligned}$$

1.4 Classification of Linear Integral Equations

(a) Fredholm Equations

Before beginning to discuss the theory of integral equations (it will be *linear* integral equations that will be the main concern of this book), it will be necessary to make some definitions and introduce a preliminary classification of linear integral equations. In what follows, ϕ will be the unknown function in the integral equation and f will be a known function.

The integral equation

$$\int_a^b K(x,y)\phi(y)\,dy = f(x) \tag{1.49}$$

is termed a Fredholm integral equation of the first kind. The integral equation

$$\phi(x) = \lambda \int_a^b K(x,y)\phi(y)\,dy + f(x) \qquad a \leqslant x \leqslant b \tag{1.50}$$

is termed a Fredholm integral equation of the second kind. It is not necessary to include λ, but it is convenient to do so from the point of view of the development of the theory. The integral equation

$$\phi(x) = \lambda \int_a^b K(x,y)\phi(y)\,dy \qquad a \leqslant x \leqslant b \tag{1.51}$$

is termed a homogeneous Fredholm integral equation of the second kind. This is an eigenvalue problem. To various eigenvalues λ_r, there will correspond eigenfunctions $\phi_r(x)$. Clearly if Eq. (1.50) has more than one solution, then the difference of the solutions will satisfy Eq. (1.51). Thus it is impossible for there to be eigenfunctions and for Eq. (1.50) to have a unique solution simultaneously for a given value of λ.

The quantity $K(x, y)$ is termed the kernel of the integral equation. In Eq. (1.49) the kernel is defined over $a \leqslant y \leqslant b$, and over the same domain of x that $f(x)$ is defined over. In Eqs. (1.50) and (1.51), the kernel is defined over the domain $a \leqslant x \leqslant b$, $a \leqslant y \leqslant b$. The integral equation is said to be singular if either a domain of definition is infinite, or if the kernel has a singularity within its region of definition. In certain cases however, the kernel is only weakly singular as the singularity may be transformed away by a change of variable. Suppose that

$$K(x,y) = H(x,y)|y-x|^{-\alpha} \tag{1.52}$$

where H is continuous and finite everywhere of interest. Then, supposing that ϕ is bounded,

$$\int_a^b K(x,y)\phi(y)\,dy = \int_a^x H(x,y)(x-y)^{-\alpha}\phi(y)\,dy + \int_x^b H(x,y)(y-x)^{-\alpha}\phi(y)\,dy$$

Consider the first integral and let $x-y=\eta^\gamma$, $x-a=\xi^\gamma$. Then

$$\int_a^x H(x,y)(x-y)^{-\alpha}\phi(y)\,dy = -\gamma\int_0^\xi H(x,x-\eta^\gamma)\phi(x-\eta^\gamma)\eta^{\gamma(1-\alpha)-1}\,d\eta \quad (1.53)$$

The integral contains a factor $\eta^{\gamma(1-\alpha)-1}$. The index $\gamma(1-\alpha)-1$ is positive if $\alpha<1$ and if $\gamma>(1-\alpha)^{-1}$. Thus, if $\alpha<1$, it is possible, by a suitable transformation, to transform the singular integral into an integral without a singularity. An exactly similar treatment can be applied to the integral over the range $x \leqslant y \leqslant b$, and so, for many purposes, it is possible to treat a weakly singular kernel as non-singular, and it will be assumed in future that this transformation has in fact taken place. It can be shown in an identical manner, for a kernel of the type (1.52), that

$$\int_a^b \{K(x,y)\}^2\,dy \quad (1.54)$$

is finite if $\alpha<\frac{1}{2}$.

It may be noted that Eqs. (1.49), (1.50) and (1.51) can be written in operational form as

$$K\phi = f \quad (1.55)$$

$$\phi = \lambda K\phi + f \quad (1.56)$$

$$\phi = \lambda K\phi \quad (1.57)$$

respectively. These representations will be helpful later in looking for solutions. It will be noted that, in Eq. (1.57), it is $\mu=\lambda^{-1}$ which is the eigenvalue of the operator K. Thus it will sometimes be convenient to write Eq. (1.57) in the form

$$K\phi = \mu\phi \quad (1.58)$$

It is, however, usual in integral equation work to refer to λ, rather than μ, as the eigenvalue, and this practice will be followed here.

If

$$K(x,y) = K(y,x) \quad (1.59)$$

the kernel is said to be symmetric.

If

$$K(x,y) = -K(y,x) \quad (1.60)$$

the kernel is said to be antisymmetric.

If

$$K(x, y) = \bar{K}(y, x) \tag{1.61}$$

the kernel is said to be Hermitian.

Other definitions similar to those of matrix algebra can be given, but the above will be sufficient for the purpose of this book. It will be seen that Eqs. (1.55), (1.56) and (1.57) could have a matrix algebra representation, and consequently many results in the theory of Fredholm integral equations will have analogies in the theory of matrix algebra.

If

$$K(x, y) = \sum_{r=1}^{n} a_r(x) b_s(y), \; n \text{ being finite} \tag{1.62}$$

the kernel K is said to be separable or degenerate.

(b) Volterra Integral Equations

If

$$K(x, y) = 0, \qquad y > x \tag{1.63}$$

the kernel is said to be of Volterra type.

The integral equation

$$\int_a^x K(x, y)\phi(y)\,dy = f(x) \qquad a \leqslant x \tag{1.64}$$

is termed a Volterra integral equation of the first type.

If $K(x, y) = K(y-x)$, the kernel is said to be of convolution form. The integral equation

$$\phi(x) = \lambda \int_a^x K(x, y)\phi(y)\,dy + f(x) \qquad a \leqslant x \tag{1.65}$$

is termed a Volterra integral equation of the second type. In general a Volterra integral equation of the first kind can usually be reduced to a Volterra integral equation of the second type. For, differentiating Eq. (1.64) with respect to x, it follows that

$$K(x, x)\phi(x) + \int_a^x \frac{\partial K(x, y)}{\partial x}\phi(y)\,dy = f'(x)$$

If $K(x, x)$ is non zero, it is possible to divide through by it, and it is clear that an associated Volterra integral equation of the second kind exists.

If $K(x, x)$ is identically zero, there is merely a different Volterra integral equation of the first kind.

The integral equation

$$\phi(x) = \lambda \int_a^x K(x, y)\phi(y)\,\mathrm{d}y \tag{1.66}$$

is termed a homogeneous Volterra integral equation of the second kind. It may be shown, if $K(x, y)$ is continuous, that the only possible continuous solution is the trivial zero solution. For, by Appendix C, if K is continuous and ψ is bounded, then

$$\int_a^b K(x, y)\psi(y)\,\mathrm{d}y$$

is continuous.

Thus ϕ, as defined by Eq. (1.66), is continuous if it is bounded. Let

$$|K(x, y)| \leqslant M \tag{1.67}$$

Let b be a number such that $(b-a)|\lambda|M < 1$. Let $a \leqslant x \leqslant b$. Then

$$(x-a)|\lambda|M < 1 \tag{1.68}$$

It follows from Eq. (1.66) that

$$|\phi(x)| \leqslant (x-a)|\lambda|M|\phi(x)| < |\phi(x)| \tag{1.69}$$

This is only possible if $\phi(x)$ is identically zero in $a \leqslant x \leqslant b$.

Example 1.5

Show that the integral equation

$$x\phi(x) = \lambda \int_0^x \exp\{x-y\}\,\phi(y)\,\mathrm{d}y, \qquad 0 \leqslant y$$

possesses a continuous spectrum of eigenvalues and that the eigensolutions are not bounded in $0 \leqslant x$.

This is a homogeneous Volterra equation of the second kind, with a singular kernel $x^{-1}\exp\{x-y\}$. The method of solution is obvious.

Let

$$\psi(y) = \exp\{-y\}\,\phi(y)$$

Then

$$x\psi(x) = \lambda \int_0^x \psi(y)\,\mathrm{d}y$$

The solution of this is obviously $\psi(x) = x^{\lambda-1}$, and so that the integral on the right hand side exists, it follows that the real part of λ must be positive.

Thus $\phi(x) = x^{\lambda-1} \exp x$, where λ is any quantity with positive real part. This gives a continuous spectrum of values for λ. Also $\phi(x)$ is not bounded for positive x.

1.5 Integro-differential Equations

It sometimes happens that physical situations give rise, when reduced to mathematical terms, to what are termed integro-differential equations. For example, the rate of deformation u of a circular plate caused by an underwater explosion is governed, in the early stages, by an equation of the form

$$\frac{\mathrm{d}u}{\mathrm{d}t} + \alpha u + \beta \int_0^t (1 - t'^2/t_1^2)^{\frac{1}{2}} u(t - t')\,\mathrm{d}t' = \gamma \exp\{-t/t_2\} \tag{1.70}$$

In this the quantity u occurs both as a derivative with respect to time and inside an integral sign. Another way in which integro-differential equations arise is as an intermediate stage in the conversion of a differential equation to an integral equation. Consider the solution of the differential equation

$$\frac{\mathrm{d}^2 y}{\mathrm{d}t^2} + \omega^2 y = f \cos \Omega t \qquad 0 \leqslant t \tag{1.71}$$

which has associated the initial conditions

$$y = 0, \qquad \frac{\mathrm{d}y}{\mathrm{d}t} = 0 \quad \text{at} \quad t = 0 \tag{1.72}$$

This can be transformed by the method of Section 1.3(a) into an integral equation. However, if Eq. (1.71) be integrated once only with respect to time, it becomes

$$\frac{\mathrm{d}y}{\mathrm{d}t} + \omega^2 \int_0^t y(t')\,\mathrm{d}t' = f\Omega^{-1} \sin \Omega t \tag{1.73}$$

only one initial condition, $y = 0$ at $t = 0$, now being necessary. Equation (1.73) contains both a derivative of y, and y within an integral sign, and is thus an integro-differential equation.

In general, integro-differential equations are amenable to treatment by both differential and integral equation methods as might be expected, and no further specific mention will be made of them. Sometimes the method of solution is obvious as in the following example.

Example 1.6

Find the solution of

$$u'(x)+\int_0^1 \exp\{x-y\}\,u(y)\,\mathrm{d}y = f(x) \qquad 0 \leqslant x \leqslant 1$$

where $u(0) = 0$ (Wales)

The equation can be rewritten as

$$u'(x)+\exp x\int_0^1 \exp(-y)u(y)\,\mathrm{d}y = f(x)$$

Let

$$g(x) = \int_0^x f(\xi)\,\mathrm{d}\xi, \qquad K = \int_0^1 \exp(-y)u(y)\,\mathrm{d}y$$

Integrating, it follows that

$$u(x)+K(\exp x-1) = g(x)$$

and

$$u(x) = g(x)+K(1-\exp x)$$

Also

$$K = \int_0^1 \exp(-y)u(y)\,\mathrm{d}y = \int_0^1 \exp(-y)g(y)\,\mathrm{d}y+K\int_0^1 \{\exp(-y)-1\}\,\mathrm{d}y$$

$$= \int_0^1 \exp(-y)g(y)\,\mathrm{d}y-K\,\mathrm{e}^{-1}$$

whence

$$K = [\mathrm{e}/(\mathrm{e}+1)]\int_0^1 \exp(-y)g(y)\,\mathrm{d}y$$

EXERCISES

1. Turn the following differential equation for y into an integral equation for f:

$$y''+y = f, \qquad y = 0 \quad \text{at} \quad x = 0, \qquad x = \pi/2$$

2. Turn the following differential equation for y into an integral equation:

$$y''-\lambda y= \cos x, \qquad y = 0 \quad \text{at} \quad x = 0, \qquad y' = 0 \quad \text{at} \quad x = 1, \qquad \lambda > 0$$

3. If $G(x, \xi) = \sinh x \sinh(\xi - a) \operatorname{cosech} a \qquad 0 \leqslant x \leqslant \xi$

$\qquad\qquad = \sinh \xi \sinh(x - a) \operatorname{cosech} a \qquad \xi \leqslant x \leqslant a$

find the differential equation and boundary conditions which are equivalent to the integral equation

$$y(x) = \int_0^a G(x, \xi) f(\xi) \,\mathrm{d}\xi$$

4. Find possible eigenvalues and eigenfunctions for the integral equation

$$x^2 \phi(x) = \lambda \int_0^x [a(y)/a(x)] \phi(y) \,\mathrm{d}y$$

5. Show that a linear differential equation with constant coefficients and arbitrary initial conditions can be transformed into a Volterra integral equation of the second kind with a convolution type kernel.

6. Verify that the integral equation

$$\phi(x) = \int_0^x t^{x-t} \phi(t) \,\mathrm{d}t \qquad 0 \leqslant x$$

has a discontinuous solution x^{x-1}.

7. Show that, if

$$\frac{\mathrm{d}}{\mathrm{d}x}\left\{p(x)\frac{\mathrm{d}y}{\mathrm{d}x}\right\} - q(x)y = f(x) \qquad a \leqslant x \leqslant b$$

and y is subject to the boundary conditions

$$\alpha_1 y(a) + \alpha_2 y'(a) + \alpha_3 y(b) + \alpha_4 y'(b) = 0$$

and

$$\beta_1 y(a) + \beta_2 y'(a) + \beta_3 y(b) + \beta_4 y'(b) = 0$$

there exists an integral equation for $f(x)$ of the form

$$y(x) = \int_a^b G(x, \xi) f(\xi) \,\mathrm{d}\xi$$

and that $G(x, \xi)$ is symmetric in x and ξ if

$$p(b) \begin{vmatrix} \alpha_1 & \alpha_2 \\ \beta_1 & \beta_2 \end{vmatrix} = p(a) \begin{vmatrix} \alpha_3 & \alpha_4 \\ \beta_3 & \beta_4 \end{vmatrix}$$

Show also that G is symmetric if

$$y(a) = y(b), \quad y'(a) = y'(b) \quad \text{and} \quad p(a) = p(b)$$

8. Show that the solution of the integral equation for $\phi(x)$

$$f(\omega) = \int_{-\infty}^{\infty} \phi(x) \cos \omega x \, dx$$

is

$$\phi(x) = \frac{1}{2\pi} \int_{-\infty}^{\infty} f(\omega) \cos \omega x \, d\omega$$

provided that the integrals exist.

9. Solve the integral equation

$$x^{\alpha} = \int_0^x (x-\xi)^{\beta} \phi(\xi) \, d\xi \qquad (\beta \text{ integral}) \; 0 \leqslant x$$

stating for what values of α and β the solution is valid. (Hint, use the result of Appendix A.)

10. Solve the integral equation of the first kind

$$\int_0^x e^{x-t} \phi(t) \, dt = f(x), \qquad f(0) = 0 \qquad 0 \leqslant x$$

by reducing it to an integral equation of the second kind.

11. Solve the integral equation

$$\int_0^x \sin \omega(x-\xi) \, \phi(\xi) \, d\xi = f(x) \qquad 0 \leqslant x$$

where $f(0) = 0, f'(0) = 0$.

2 Fredholm Equations

2.1 Analogies with Matrix Algebra

Consider the three integral equations

$$f(x) = \int K(x, y)\phi(y)\mathrm{d}y \tag{2.1}$$

$$\phi(x) = \lambda\int K(x, y)\phi(y)\mathrm{d}y \tag{2.2}$$

$$\phi(x) = \lambda\int K(x, y)\phi(y)\mathrm{d}y + f(x) \tag{2.3}$$

The ranges of integration and definition of the functions involved are all $a \leqslant x,\ y \leqslant b$. Limits of integration will not be mentioned unless necessary. Before actually discussing the solution of these equations, it may be helpful to obtain simple approximations to them and to discuss the approximations. In this manner, it will be possible to get an idea of what the properties of the integral equations are, although in general these properties will be indicated rather than proved. It will be assumed that the equations are non-singular.

Let n be an integer and let p and q be positive integers less than n. Let

$$h = \frac{b-a}{n} > 0$$

As n tends to infinity and h tends to zero, it would be reasonable to expect the approximation to become better and better.

$$x_p = a+(p-\tfrac{1}{2})h, y_q = a+(q-\tfrac{1}{2})h$$
$$f(x_p) = f_p, \phi(x_p) = \phi_p, K(x_p, y_q) = k_{pq}$$

Now $h\sum\limits_{q=1}^{n} g_q$ is an approximation to $\int g(y)\mathrm{d}y$ and so the sets of equations for ϕ_q

$$f_p = h\sum_{q=1}^{n} k_{pq}\phi_q \qquad 1 \leqslant p \leqslant n \tag{2.4}$$

$$\phi_p = \lambda h\sum_{q=1}^{n} k_{pq}\phi_q \qquad 1 \leqslant p \leqslant n \tag{2.5}$$

$$\phi_p = \lambda h \sum_{q=1}^{n} k_{pq}\phi_p + f_p \qquad 1 \leqslant p \leqslant n \tag{2.6}$$

are respectively approximations to the integral equations (2.1), (2.2) and (2.3) for $\phi(y)$. Equations (2.4), (2.5) and (2.6) may be rewritten respectively in the obvious matrix form

$$F = hK\Phi \tag{2.7}$$

$$\Phi = \lambda hK\Phi \tag{2.8}$$

$$\Phi = \lambda hK\Phi + F \tag{2.9}$$

where K is the $n \times n$ square matrix with elements k_{pq}, F and Φ are column matrices with n elements f_p and ϕ_p respectively.

Consider now the behaviour of these matrix equations.[2] Equation (2.7) has a unique solution

$$\Phi = (hK)^{-1}F$$

provided that K is a non-singular matrix. If however K is non-singular, the rank of K is less than its order and some of its rows are linearly independent of the others. If the same relationship holds between the corresponding elements of Φ, there will be an infinity of non-unique solutions.[3] If this is not the case, the equations are inconsistent and there is not any solution. Thus there is an indication that Eq. (2.1) may be soluble with a unique solution, or an infinity of solutions or it may not have any solution.

Consider now Eq. (2.8), which may be rewritten as

$$hK\Phi = \mu\Phi, \quad \mu = \lambda^{-1}$$

If K is non-singular then this equation has n eigenvectors Φ_r and n non-zero eigenvalues associated with it. It may be assumed that the eigenvalues are all different. Where they are not, suitable modifications may be made to the theory. If the matrix K is singular and has rank $m < n$, there will be $n-m$ eigenvectors corresponding to a zero eigenvalue. It should be noted that the eigenvectors Ψ given by the solutions of

$$\Psi^T hK = \mu\Psi^T$$

where the superscript T denotes a transpose, will not in general be the same as the Φ, unless the matrix K is symmetric. However, the eigenvalues will always be the same.

Some orthogonality relations may be proved as follows: Suppose that the eigenvectors Ψ_r, Φ_s correspond to non-zero, non-equal eigenvalues μ_r, μ_s respectively,

$$\Psi_r^T h^2 K\Phi_s = \Psi_r^T h\mu_s\Phi_s = (\mu_s/\mu_r)\Psi_r^T h^2 K\Phi_s$$

which is only possible if

$$\Psi_r^T h^2 K \Phi_s = \Psi_r^T h^2 \Phi_s \tag{2.10}$$

vanishes. By the usual orthogonalization process, this result can be made to hold in the other case when the eigenvalues are equal. Furthermore, it is always possible, by a change of scale, to make

$$h\Psi_r^T \Psi_r = h\Phi_s^T \Phi_s = 1 \tag{2.11}$$

When this normalization process is carried out, it is clear that

$$\Psi_s^T hK\Phi_s = \mu_s = \lambda_s^{-1} \tag{2.12}$$

Suppose Q is an arbitrary column matrix with n elements.

Let

$$Q = \sum_{s=1}^{n} q_s \Phi_s$$

Then

$$h\Psi_r^T Q = \sum_{s=1}^{n} q_s h\Psi_r^T \Phi_s = q_r$$

and so

$$Q = \sum_{s=1}^{n} h\Psi_s^T Q \Phi_s$$

Alternatively

$$q_r = \lambda_r h \Psi_r^T KQ$$

Consider now the solution of the set of equations (2.9)

$$\Phi = \lambda hK\Phi + F$$

If λ is not an eigenvalue of the set of equations

$$\Phi = \lambda hK\Phi$$

that is

$$D_0(\lambda) = |I - \lambda hK| \neq 0 \tag{2.13}$$

(I is the unit matrix of order n) then the set of equations (2.6) has a unique solution.

It can be seen that $D_0(\lambda)$ is a determinant of order n and that it can be expanded as a polynomial of the nth degree in λ and that the constant term is unity. $D_0(\lambda)$ must also vanish when λ is an eigenvalue of the system defined by Eq. (2.8).

Thus

$$D_0(\lambda) = \prod_{s=1}^{n} (1-\lambda/\lambda_s) = \prod_{s=1}^{n} (1-\mu_s/\mu)$$

where the λ_s are the eigenvalues of the system. It will be convenient to assume

$$|\lambda_1| \leqslant |\lambda_2| \leqslant \dots \quad \leqslant |\lambda_n|$$

Correspondingly

$$|\mu_1| \geqslant |\mu_2| \geqslant \dots \quad \geqslant |\mu_n|$$

and this convention will be followed hereafter in this book.

Look for a solution of the form

$$\Phi = \sum_{r=1}^{n} y_r \Phi_r$$

where the Φ_r are the eigenvectors of the homogeneous system. Now

$$F = \sum_{r=1}^{n} \xi_r \Phi_r \text{ where } \xi_r = \Psi_r^T F$$

and so

$$\sum_{r=1}^{n} y_r \Phi_r = \lambda h \sum_{r=1}^{n} y_r K \Phi_r + \sum_{r=1}^{n} \xi_r \Phi_r$$

whence

$$y_r \left(1 - \frac{\lambda}{\lambda_r}\right) = \xi_r$$

and so

$$y_r = \frac{\xi_r}{(1-\lambda/\lambda_r)} = \xi_r + \frac{\lambda \xi_r}{\lambda_r - \lambda} = \xi_r + \frac{\lambda \lambda_r \Psi_r^T hKF}{\lambda_r - \lambda}$$

Thus

$$\Phi = F + \lambda \sum_{r=1}^{n} \frac{\Psi_r^T hKF}{1-\lambda/\lambda_r} \Phi_r = F + \lambda \{D_0(\lambda)\}^{-1} \sum_{r=1}^{n} \prod_{\substack{s=1 \\ s \neq n}}^{n} (1-\lambda/\lambda_s) \Psi_r^T hKF \Phi_r \tag{2.14}$$

It is fairly easy to see that this can be written in the form

$$\phi_p = f_p - \lambda \{D_0(\lambda)\}^{-1} \sum_{q=1}^{n} hd_{pq}(\lambda) f_q \tag{2.15}$$

where the $d_{pq}(\lambda)$ are polynomials of order $n-1$ in λ. If λ is an eigenvalue, say λ_t, there will be an infinity in the term $\lambda\xi_t/(\lambda_t-\lambda)$ unless $\xi_t = \Psi_t^T F$ vanishes, when there will be an indeterminacy, and the solution will be indefinite by a quantity $C\Phi_t$ where C is an arbitrary scalar.

There is another way of solving the equation

$$\Phi = \lambda hK\Phi + F$$

Consider the sequence of vectors defined by

$$\Phi^{(0)} = F$$
$$\Phi^{(r+1)} = \lambda hK\Phi^{(r)} + F$$

It is not difficult to see that

$$\Phi^{(r)} = \sum_{s=0}^{r} \lambda^s h^s K^s F$$

Now

$$F + \lambda hK\Phi^{(r)} - \Phi^{(r)} = \lambda^{r+1} h^{r+1} K^{r+1} F$$

and if this last quantity tends to zero then the sequence $\Phi^{(r)}$ will tend to a solution of

$$F + \lambda hK\Phi - \Phi = 0$$

Also

$$F = \sum_{p=1}^{n} \xi_p \Phi_p$$

and so

$$\lambda^{r+1} h^{r+1} K^{r+1} F = \sum_{p=1}^{n} \left(\frac{\lambda}{\lambda_p}\right)^{r+1} \Phi_p$$

and this will tend to zero if

$$|\lambda| < |\lambda_1| \leqslant |\lambda_2| \leqslant \dots \quad |\lambda_n|$$

Thus, if $|\lambda| < |\lambda_1|$, a solution is given by

$$\Phi = \sum_{s=0}^{\infty} \lambda^s h^s K^s F \tag{2.16}$$

which is formally equivalent to the relation

$$\Phi = (I - \lambda hK)^{-1} F$$

An alternative expression is

$$\Phi = F - \lambda hRF \tag{2.17}$$

where

$$R = -K(I-\lambda hK)^{-1} = -\sum_{s=0}^{\infty} \lambda^s h^s K^{s+1} F \tag{2.18}$$

In order to consider how the solutions of the integral equations may be expected to behave, the following interpretation may be used. The elements of the column matrix defined by

$$hAB$$

where A is a square matrix and B is a column matrix are

$$h \sum_{q=1}^{n} a_{pq} b_q = h \sum_{q=1}^{n} A(x_p, y_q) B(y_q) \tag{2.19}$$

a_{pq} being the value at x_p, y_q of the function $A(x, y)$ and b_q being that of the function $B(y)$ at y_q, $A(x, y)$, $B(y)$ being some suitable functions. Thus expression (2.19) is an approximation to

$$\int A(x_p, y) B(y) \mathrm{d}y$$

From these remarks and what has gone before, the following properties of the solutions of integral equations are suggested:

(a) The equation

$$f(x) = \int K(x, y) \phi(y) \mathrm{d}y$$

has a unique solution provided that there do not exist any functions $\Psi(y)$ such that

$$\int K(x, y) \Psi(y) \mathrm{d}y = 0 \tag{2.20}$$

(b) There exists an infinite number of eigenfunctions $\Phi_s(y)$, $\Psi_s(x)$ and eigenvalues λ_s associated with K such that

$$\lambda_s \int K(x, y) \Phi_s(y) \mathrm{d}y = \Phi_s(x) \tag{2.21}$$

$$\lambda_s \int \Psi_s(x) K(x, y) \mathrm{d}x = \Psi_s(y) \tag{2.22}$$

Φ_s and Ψ_s are the same if K is symmetrical in x and y, and Φ_s and Ψ_s can be normalized so that

$$\int \Psi_r(x) \Phi_s(x) \mathrm{d}x = \delta_{rs} \tag{2.23}$$

$$\int\int \Psi_r(x)K(x, y)\Phi_s(y)\,dx\,dy = \lambda_s^{-1}\delta_{rs} = \mu_s\delta_{rs} \qquad (2.24)$$

where $\delta_{rs} = 1, \quad r = s, = 0 \quad r \neq s$.

It is possible that some of the μ_s may be zero.

Equations (2.10), (2.11) and (2.12) are approximations to Eqs. (2.23) and (2.24).

(c) The integral equation

$$\phi(x) = \lambda\int K(x, y)\phi(y)\,dy + f(x)$$

has a unique solution of the form

$$\phi(x) = f(x) - \lambda\int R(x, y; \lambda)f(y)\,dy \qquad (2.25)$$

unless λ is an eigenvalue. $R(x, y; \lambda)$ is of the form $D(x, y; \lambda)/D_0(\lambda)$ where

$$D_0(\lambda) = \prod_{r=1}^{\infty} (1 - \lambda/\lambda_r)$$

where the λ_r are the eigenvalues. $D(x, y; \lambda)$ can be expressed as a power series in λ. Equation (2.15) is an approximation to Eq. (2.25).

If λ is an eigenvalue λ_t, then a solution exists only if

$$\int f(x)\Psi_t(x)\,dx = 0 \qquad (2.26)$$

and in this case the solution is indefinite by an amount $C\Phi_t(x)$ where C is an arbitrary constant.

An alternative expression for the solution which is convergent when $|\lambda| < |\lambda_1|$, where λ_1 is the smallest eigenvalue is

$$\phi(x) = f(x) + \int \sum_{n=1}^{\infty} \lambda^n K_n(x, y) f(y)\,dy \qquad (2.27)$$

where $K_n(x, y)$ is defined by

$$K_n(x, y) = \int K_{n-1}(x, z)K(z, y)\,dz \qquad n > 1$$
$$= K(x, y) \qquad n = 1$$

Equation (2.16) is the approximation to Eq. (2.27).

2.2 Degenerate Kernels

Consider the kernel of the form

$$K(x, y) = \sum_{p=1}^{n} a_p(x) b_p(y)$$

where n is finite, and the a_r and b_s form linearly independent sets. (If this is not the case, the number of terms may be reduced.) A kernel of this character is termed a degenerate kernel.

Consider now the integral equation of the first kind:

$$f(x) = \int K(x, y)\phi(y)\mathrm{d}y \tag{2.28}$$

$$= \sum_{p=1}^{n} a_p(x) \int b_p(y)\phi(y)\mathrm{d}y$$

Two statements can be made immediately:

(a) No solution exists, unless $f(x)$ can be written in the form

$$\sum_{p=1}^{n} f_p a_p(x)$$

This is essential for the equation to be self-consistent.

(b) The solution is indefinite by any function $\psi(y)$ which is orthogonal to all of the $b_p(y)$ over the range of integration. Such a function can always be constructed when n is finite. Thus, it is necessary to verify that the equation is self-consistent, and when looking for a solution it will be convenient to look only for the simplest.

Example 2.1

The integral equation

$$\exp 2x = \int_0^{\pi} \sin(x+y)\phi(y)\mathrm{d}y \qquad 0 \leqslant x \leqslant \pi$$

is not self-consistent and so does not have a solution. This is because

$$\int_{-\pi}^{\pi} \sin(x+y)\phi(y)\mathrm{d}y = \sin x \int_{-\pi}^{\pi} \cos y\, \phi(y)\mathrm{d}y + \cos x \int_{-\pi}^{\pi} \sin y\, \phi(y)\mathrm{d}y$$

which is of the form

$$A \sin x + B \cos x$$

Example 2.2

The solution of the integral equation

$$3\sin x + 2\cos x = \int_{-\pi}^{\pi} \sin(x+y)\phi(y)\mathrm{d}y \qquad -\pi \leqslant x \leqslant \pi$$

is indefinite by a quantity of the form

$$\psi(y) = C_0 + \sum_{n=2}^{\infty} (C_n \cos ny + d_n \sin ny)$$

because

$$\int_{-\pi}^{\pi} \psi(y)\sin(x+y)\mathrm{d}y = 0$$

The general method of solution is as follows. Look for a solution of the form

$$\phi(y) = \sum_{q=1}^{n} \phi_q b_q(y)$$

If it exists, it will be a solution and it is possible to add the $\psi(y)$ to it. This process is similar to the idea of the particular integral and complementary function in differential equation theory. The solution proceeds as follows:

$$\sum_{p=1}^{n} f_p a_p(x) = \sum_{p=1}^{n} a_p(x) \int b_p(y) \sum_{q=1}^{n} \phi_q b_q(y)\,\mathrm{d}y = \sum_{p=1}^{n} a_p(x) \sum_{q=1}^{n} \beta_{pq}\phi_q$$

where

$$\beta_{pq} = \int b_p(y) b_q(y)\,\mathrm{d}y$$

and so the ϕ_s are defined by

$$f_p = \sum_{q=1}^{n} \beta_{pq}\phi_q \qquad 1 \leqslant r \leqslant n$$

Because the b_p are linearly independent the determinant $|\beta_{pq}|$ does not vanish and the ϕ_q can be found uniquely. This solution may be termed a particular solution and the $\psi(y)$ a complementary function.

Example 2.3

Solve the integral equation

$$3\sin x + 2\cos x = \int_{-\pi}^{\pi} \sin(x+y)\phi(y)\mathrm{d}y \qquad -\pi \leqslant x \leqslant \pi$$

Now $\sin(x+y) = \sin x \cos y + \cos x \sin y$ and there is consistency. Look for a solution of the form

$$\phi(y) = A\cos y + B\sin y$$

$$\begin{aligned}\int_{-\pi}^{\pi} \sin(x+y)\phi(y)\mathrm{d}y &= \sin x \int_{-\pi}^{\pi} \cos y(A\cos y + B\sin y)\mathrm{d}y \\ &\quad + \cos x \int_{-\pi}^{\pi} \sin y(A\cos y + B\sin y)\mathrm{d}y \\ &= \pi A \sin x + \pi B \cos x\end{aligned}$$

whence $A = 3/\pi$, $B = 2/\pi$ and the particular solution is

$$\phi(y) = (3\cos y + 2\sin y)/\pi$$

with a complementary function

$$C_0 + \sum_{n=2}^{\infty} (C_n \cos ny + d_n \sin ny)$$

as indicated in Example 2.2.

Example 2.4

Solve the integral equation

$$3x^2 + 4x = \int_{-1}^{+1} (6x^2y + 4xy^2)\phi(y)\mathrm{d}y \qquad -1 \leqslant x \leqslant 1$$

Equating powers of x, it follows that

$$1 = 2\int_{-1}^{+1} y\phi(y)\mathrm{d}y$$

$$1 = \int_{-1}^{+1} y^2\phi(y)\mathrm{d}y$$

There are no other powers of x and so there is consistency. It is not easy to see an obvious form for the complementary function, so consider expansions in terms of Legendre polynomials. These are convenient because they are orthogonal over $-1 \leqslant y \leqslant 1$. Now $y = P_1(y)$

$$y^2 = \frac{2P_2(y) + P_0(y)}{3}$$

Now because of the orthogonality properties of Legendre polynomials, the set of functions

$$P_0(y)-\tfrac{2}{5}P_2(y),\ P_n(y) \qquad n \geqslant 3$$

are orthogonal over $-1 \leqslant y \leqslant 1$ to y and y^2. Thus the complementary function will be

$$C_0(P_0(y)-\tfrac{2}{5}P_2(y))+\sum_{n=3}^{\infty} C_n P_n(y)$$

The particular solution follows easily because y and y^2 are orthogonal.

Put

$$\phi(y) = \phi_1 y+\phi_2(y^2/2)$$

Then

$$1 = 2\int_{-1}^{+1} y[\phi_1 y+\phi_2 y^2/2]\mathrm{d}y = 2\phi_1\int_{-1}^{+1} y^2\,\mathrm{d}y$$

$$1 = 2\int_{-1}^{+1} y^2[\phi_1 y+\phi_2 y^2/2]\mathrm{d}y = \phi_2\int_{-1}^{+1} y^4\,\mathrm{d}y$$

whence the particular solution

$$\phi(y) = \frac{3y}{4}+\frac{5y^2}{2}\text{follows}$$

Consider now possible eigenfunctions and eigenvectors associated with the equation

$$\phi(y) = \lambda\int\sum_{p=1}^{n} a_p(x)b_p(y)\phi(y)\mathrm{d}y \tag{2.29}$$

If Eq. (2.29) is rewritten as

$$\mu\phi(y) = \int\sum_{p=1}^{n} a_p(x)b_p(x)\phi(y)\mathrm{d}y \tag{2.30}$$

it is clear that it is satisfied by any function $\phi(y)$ such that

$$\int b_p(y)\phi(y)\mathrm{d}y$$

is zero and μ is zero. It is sometimes convenient to regard such functions also as eigenfunctions, but in general they will be ignored.

Any eigenfunction must be of the form

$$\Phi(x) = \sum_{p=1}^{n} \phi_p a_p(x)$$

and so

$$\sum_{p=1}^{n} \phi_p a_p(x) = \lambda \sum_{p=1}^{n} a_p(x) \int b_p(y) \sum_{q=1}^{n} \phi_q a_q(y) \mathrm{d}y$$

whence

$$\phi_p = \sum_{q=1}^{n} \phi_q k_{pq}$$

$$k_{pq} = \int b_p(y) a_q(y) \mathrm{d}y$$

The eigenvalues and eigenfunctions follow by the usual processes. If, however, all the b_r are orthogonal to all the a_q, k_{pq} is zero for all p and q and this again will correspond to zero μ, the ϕ_p being arbitrary.

Example 2.5

Find the eigenvalues and eigenfunctions of the system defined by

$$\phi(x) = \int_0^1 (1+xt)\phi(t)\,\mathrm{d}t \qquad 0 \leqslant x \leqslant 1$$

Let

$$\Phi(x) = \phi_0 + \phi_1 x = \int_0^1 (1+xt)(\phi_0 + \phi_1 t)\mathrm{d}t$$
$$= (\phi_0 + \phi_1/2) + (\phi_0/2 + \phi_1/3)x$$

whence

$$(\lambda - 1)\phi_0 + \lambda\phi_1/2 = 0$$
$$\lambda\phi_0/2 + (\lambda/3 - 1)\phi_1 = 0$$
$$(\lambda - 1)(\lambda/3 - 1) = \lambda^2/4,\ \lambda = 8 \pm \sqrt{52}$$

and

$$\phi_1 : \phi_0 = -(7 \pm \sqrt{52}) : (8 \pm \sqrt{52})$$

Example 2.6

Find the eigenvalues and eigenfunctions of the system defined by

$$\phi(x) = \int_0^1 (4x-3)t^2\phi(t)\,\mathrm{d}t \qquad 0 \leqslant x \leqslant 1$$

$\Phi(x) = k(4x-3)$ is the only possible eigenfunction. However

$$k(4x-3) = \int_0^1 (4x-3)(4t^3-3t^2)k\,\mathrm{d}t = 0$$

Thus $(4x-3)$, the eigenfunction, corresponds to a zero value of μ; that is, there is no value of λ for which the equation can be satisfied.

Consider now the solution of the integral equation

$$\phi(x) = \lambda \int \sum_{r=1}^{n} a_p(x)b_p(y)\phi(y)\,\mathrm{d}y + f(x) \tag{2.31}$$

Any solution will obviously be of the form

$$\phi(x) = \lambda \sum_{p=1}^{n} \phi_p a_p(x) + f(x)$$

$$= \lambda \int \sum_{p=1}^{n} a_p(x)b_p(y)\left\{\lambda \sum_{q=1}^{n} \phi_q a_q(y) + f(y)\right\}\mathrm{d}y + f(x)$$

whence, on equating terms in $a_p(x)$, it follows that

$$\phi_p = \lambda \sum_{q=1}^{n} k_{pq}\phi_q + f_p \qquad 1 \leqslant p \leqslant n \tag{2.32}$$

The solution of this is completely analogous with that of Eq. (2.6). Equation (2.32) may be written as

$$\phi_p - \lambda \sum_{q=1}^{n} k_{pq}\phi_q = f_p$$

and it follows that

$$\phi_q = \{D_0(\lambda)\}^{-1} \sum_{p=1}^{n} d_{pq}(\lambda) f_p$$

where $D_0(\lambda)$ is defined as previously and $d_{pq}(\lambda)$ is the cofactor of the p, q element of the determinant

$$|\delta_{pq} - \lambda k_{pq}|$$

Thus

$$\phi(x) = f(x) + \lambda\{D_0(\lambda)\}^{-1} \sum_{p=1}^{n} a_p(x)d_{pq}(\lambda) \int \sum_{q=1}^{n} b_q(y)f(y)\,\mathrm{d}y$$

$$= f(x) - \lambda\{D_0(\lambda)\}^{-1} \int D(x, y : \lambda) f(y)\,\mathrm{d}y$$

where $D(x, y; \lambda)$ can be seen, using the definition of d_{rs}, to be

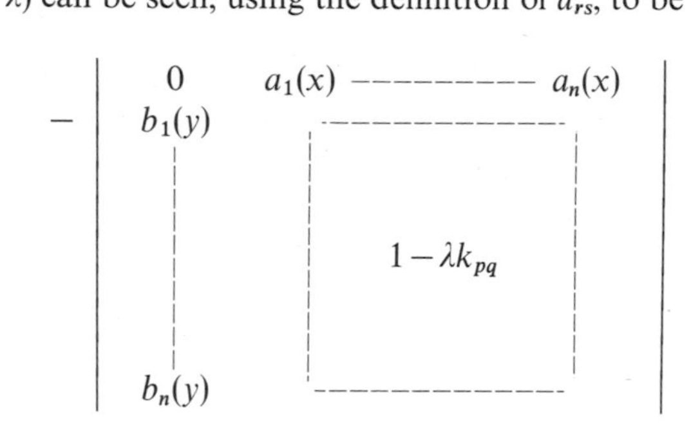

$D(x, y:\lambda)$ is a polynomial of degree n in λ. An alternative method of writing the solution is

$$\phi(x) = f(x) - \lambda \int R(x, y:\lambda) f(y)\,\mathrm{d}y$$

R is termed the resolvent kernel and is the ratio of two polynomials of degree n in λ.

It may be remarked that, to the transposed kernel

$$K^T(x, y) = K(y, x) = \sum_{p=1}^{n} a_p(y) b_p(x)$$

there correspond the same eigenvalues and hence the same $D_0(\lambda)$. The eigenfunctions Ψ_r need not, however, be the same as the corresponding Φ_r. If K is symmetrical they will be the same of course. There exist orthogonality properties which can be proved

$$\int \Psi_p(x)\Phi_q(x)\,\mathrm{d}x = \lambda_q \int \Psi_p(x) \int K(x, y)\,\Phi_q(y)\,\mathrm{d}y\,\mathrm{d}x$$

$$= (\lambda_q/\lambda_p) \int \Psi_p(y)\Phi_q(y)\,\mathrm{d}y$$

Thus if λ_q and λ_p are unequal

$$\int \Psi_p(x)\Phi_q(x)\,\mathrm{d}x$$

and

$$\iint \Psi_p(x) K(x, y) \Phi_q(y)\,\mathrm{d}x\,\mathrm{d}y$$

must both vanish. It was suggested in Section 2.1 that this was the case, and it has now been proved. The result is in fact true for any kernel, not merely degenerate ones.

When λ in Eq. (2.31) is one of the eigenvalues of the kernel, say λ_t, and there are p eigenfunctions $\Phi_{t\alpha}(x)$ corresponding to this eigenvalue, the solution will be arbitrary by a function

$$\sum_{\alpha=1}^{p} u_\alpha \Phi_{t\alpha}(x)$$

where the u_α are arbitrary constants. For consistency the following results must hold.

$$\int f(x)\Psi_{t\alpha}(x)\,dx = \int \phi(x)\Psi_{t\alpha}(x)\,dx - \lambda_t \iint \Psi_{t\alpha}(x)K(x,y)\phi(y)\,dx\,dy$$
$$= 0$$

The remainder of the solution follows in the usual way.

An alternative process for defining $R(x, y:\lambda)$ is as follows. It is analogous with the iterative solution of the matrix equation (2.9). Define a sequence of iterated kernels.

$$K_1(x,y) = K(x,y)$$
$$K_{n+1}(x,y) = \int K(x,t)K_n(t,y)\,dt$$

It can be proved that

$$K_{m+n}(x,y) = \int K_m(x,t)K_n(t,y)\,dt$$

Consider a set of functions defined by

$$\phi^{(0)}(x) = f(x)$$
$$\phi^{(n+1)}(x) = f(x) + \lambda \int K(x,y)\phi^{(n)}(y)\,dy$$

It can be seen that

$$\phi^{(n)}(x) = f(x) + \lambda \int K(x,y)\phi(y)\,dy$$
$$+ \ldots + \lambda^n \int K(x,t_1)\,dt_1 \int K(t_1,t_2)\,dt_2 \ldots \int K(t_{n-1},y)f(y)\,dy$$
$$= f(x) + \sum_{r=1}^{n} \lambda^n \int K_n(x,y)f(y)\,dy$$

This series is termed the Neumann Series.

It can be shown (see Appendix D) that a sufficient condition for the convergence of the sequence $\{\phi^{(n)}(x)\}$ is that

$$|\lambda| < \left\{ \iint |K(x, y)|^2 \, dx \, dy \right\}^{-\frac{1}{2}} \tag{2.33}$$

also

$$\phi^{(n)}(x) - \lambda \int K(x, y)\phi^{(n)}(y)\, dy - f(x) = -\lambda^{n+1} \int K_{n+1}(x, y) f(y)\, dy \tag{2.34}$$

and it is shown in Appendix D that, under the same conditions, the quantity on the right hand side of Eq. (2.34) tends to zero. Thus it follows that if λ is not an eigenvalue there is a unique solution

$$\phi(x) = f(x) + \int \sum_{n=1}^{\infty} \lambda^n K_n(x, y) f(y)\, dy \tag{2.35}$$

with a resolvent kernel

$$R(x, y : \lambda) = -\sum_{n=1}^{\infty} \lambda^{n-1} K_n(x, y) \tag{2.36}$$

R has an infinity at every eigenvalue and so the radius of convergence is less than or equal to $|\lambda_1|$, λ_1 being the eigenvalue with smallest absolute value. The convergence criterion is however a sufficient one and so

$$\left\{ \iint |K(x, y)|^2 \, dx \, dy \right\}^{-\frac{1}{2}} < |\lambda_1|$$

It will be noticed that the theory is also applicable for general kernels as nowhere has the actual nature of the kernel been taken into account. All that is necessary is that the convergence condition (2.33) holds.

Example 2.7

Solve the integral equation

$$\phi(x) = \lambda \int_0^1 (1 + xt)\phi(t)\, dt + f(x)$$

(the kernel here is the same as that of Example 2.5).

Let

$$\begin{aligned} \phi(x) &= \phi_0 + \phi_1 x + f(x) \\ &= \lambda \int_0^1 (1 + xt)[\phi_0 + \phi_1 t + f(t)]\, dt + f(x) \\ &= \lambda(\phi_0 + \phi_1/2 + f_0) + \lambda x(\phi_0/2 + \phi_1/3 + f_1) + f(x) \end{aligned}$$

where

$$f_r = \int_0^1 t^r f(t)\,\mathrm{d}t$$

Equating powers of x and solving for ϕ_0 and ϕ_1, it follows that

$$\phi_0[\lambda^2 - 16\lambda + 12] = [-4\lambda(\lambda-3)f_0 + 6\lambda^2 f_1]$$

$$\phi_1[\lambda^2 - 16\lambda + 12] = [6\lambda^2 f_0 - 12\lambda(\lambda-1)f_1]$$

The eigenvalues are of course given by the roots of the equation

$$\lambda^2 - 16\lambda + 12 = 0$$

If λ is one of the eigenvalues, say $8+\sqrt{52}$, a solution is possible only if

$$0 = \int_0^1 f(x)\{8+\sqrt{52}-(7+\sqrt{52})x\}\,\mathrm{d}x$$

and the solution is indefinite by an arbitrary multiple of

$$8+\sqrt{52}-(7+\sqrt{52})x$$

Example 2.8

Find the solution of the integral equation

$$\phi(x) = \lambda \int_0^1 \exp\{k(x-y)\}\phi(y)\,\mathrm{d}y + f(x) \qquad 0 \leqslant x \leqslant 1$$

The solution can of course be obtained by using the procedure of putting

$$\phi(x) = C\exp kx + f(x)$$

However, the iterated kernel method is particularly convenient, for, if

$$K(x,y) = \exp\{k(x-y)\}$$

$$K_2(x,y) = \int_0^1 K(x,z)K(z,y)\,\mathrm{d}z$$

$$= \int_0^1 \exp\{k(x-z+z-y)\}\,\mathrm{d}z = \exp\{k(x-y)\}$$

Similarly all the iterated kernels are

$$\exp\{k(x-y)\}$$

and so

$$R(x,y:\lambda) = -\sum_{n=1}^{\infty} \lambda^{n-1}K_n(x,y) = -\exp\{k(x-y)\}/(1-\lambda)$$

and

$$\phi(x) = f(x) + \lambda/(1-\lambda)\exp(kx)\int_0^1 f(y)\exp(-ky)\,\mathrm{d}y$$

If λ is unity, the solution given is invalid, and a solution is possible only if

$$\int_0^1 f(y)\exp(-ky)\,\mathrm{d}y = 0$$

In this case, the solution is

$$\phi(x) = f(x) + C\exp kx$$

C being arbitrary

Example 2.9

Prove that

$$K(x, y) + R(x, y; \lambda) = \lambda\int K(x, z)R(z, y:\lambda)\,\mathrm{d}z$$
$$= \lambda\int R(x, z; \lambda)K(x, y)\,\mathrm{d}z$$

Now

$$K(x, z)R(z, y; \lambda) = -\int\sum_{n=1}^{\infty} K(x, z)\lambda^n K_n(z, y)\,\mathrm{d}z$$

Because the series is uniformly convergent, it is possible to integrate term by term, and the series becomes

$$-\sum_{n=1}^{\infty}\lambda^n K_{n+1}(x, y) = -\sum_{n=2}^{\infty}\lambda^{n-1}K_n(x, y)$$
$$= R(x, y) + K(x, y)$$

Now

$$\int K(x, z)R(z, y; \lambda)\,\mathrm{d}z = \int R(x, z; \lambda)K(z, y)\,\mathrm{d}z$$

as

$$\int K(x, z)K_n(z, y)\,\mathrm{d}z = \int K(z, y)K_n(x, z)\,\mathrm{d}z$$

2.3 Hermitian and Symmetric Kernels

Two particular types of kernel are of considerable importance in the theory of Fredholm integral equations. These are the Hermitian kernel and the symmetric kernel. A kernel is said to be Hermitian if it has the property

$$K(y, x) = \bar{K}(x, y)$$

where the bar denotes the complex conjugate.

A kernel is said to be symmetric, if it has the property

$$K(y, x) = K(x, y)$$

It is clear that a real Hermitian kernel is a symmetric kernel. Consequently all theorems proved for Hermitian kernels will hold for real symmetric kernels, and so in what follows it will be assumed, unless otherwise mentioned, that all kernels are Hermitian. The examples and exercises will however deal with symmetrical kernels, as these are easier to consider for illustrative purposes. As the theory applies in the more general case, and the analysis is no more difficult, the wider class of kernels is considered. A number of theorems follow almost immediately.

1. If the eigenvalues exist, they are real.

For, if

$$\Phi(x) = \lambda \int K(x, y)\Phi(y)\,dy$$

$$\int \Phi(x)\bar{\Phi}(x)\,dx = \lambda \int\int \bar{\Phi}(x)K(x, y)\Phi(y)\,dx\,dy$$

also

$$\int \bar{\Phi}(x)\Phi(x)\,dx = \bar{\lambda} \int\int \Phi(x)\bar{K}(x, y)\bar{\Phi}(y)\,dx\,dy$$

$$= \bar{\lambda} \int\int \Phi(x)K(y, x)\bar{\Phi}(y)\,dx\,dy$$

$$= \bar{\lambda} \int\int \Phi(y)K(x, y)\bar{\Phi}(x)\,dx\,dy$$

whence

$$\lambda = \bar{\lambda}$$

Thus λ is real.

2. If a kernel is Hermitian, its iterates are also Hermitian.
Let

$$K_{\alpha+\beta}(x, y) = \int K_\alpha(x, z) K_\beta(z, y)\, dz$$

where K_α and K_β are Hermitian.
Then

$$\begin{aligned}\bar{K}_{\alpha+\beta}(y, x) &= \int \bar{K}_\alpha(y, z) \bar{K}_\beta(z, x)\, dz \\ &= \int K_\alpha(z, y) K_\beta(x, z)\, dz \\ &= K_{\beta+\alpha}(x, y) = K_{\alpha+\beta}(x, y)\end{aligned}$$

and the result follows.

3. There exist orthogonality relationships between eigenfunctions associated with different eigenvalues. Suppose that $\Phi_r(x)$, $\Phi_s(x)$ are eigenfunctions associated with different eigenvalues λ_r and λ_s. It will be convenient to assume that they are normalized so that

$$\int |\Phi_r(x)|^2\, dx = 1 \tag{2.37}$$

Now

$$\Phi_r(x) = \lambda_r \int K(x, y) \Phi_r(y)\, dy$$

and

$$\begin{aligned}\bar{\Phi}_r(x) &= \lambda_r \int \bar{K}(x, y) \bar{\Phi}_r(y)\, dy \\ &= \lambda_r \int K(y, x) \bar{\Phi}_r(y)\, dy\end{aligned}$$

Thus

$$\begin{aligned}\int \bar{\Phi}_r(x) \Phi_s(x)\, dx &= \lambda_r \iint \Phi_s(x) K(y, x) \bar{\Phi}_r(y)\, dx\, dy \\ &= \lambda_r \iint \bar{\Phi}_r(x) K(x, y) \Phi_s(y)\, dx\, dy \\ &= \frac{\lambda_r}{\lambda_s} \int \bar{\Phi}_r(x) \Phi_s(x)\, dx\end{aligned}$$

If λ_r and λ_s are unequal, it follows that

$$\int \bar{\Phi}_r(x)\Phi_s(x)\,\mathrm{d}x = 0 \tag{2.38}$$

and

$$\iint \bar{\Phi}_r(x)K(x, y)\Phi_s(y)\,\mathrm{d}x\,\mathrm{d}y = 0 \tag{2.39}$$

If there are two or more eigenfunctions corresponding to one eigenvalue then by using suitable linear combinations of them, it is possible to arrange for the satisfaction of Eqs. (2.38) and (2.39) in the usual manner.

Also, using Eq. (2.37), it follows that

$$1 = \int \bar{\Phi}_r(x)\Phi_r(x)\,\mathrm{d}x = \lambda_r \iint \bar{\Phi}_r(x)K(x, y)\Phi_r(y) \tag{2.40}$$

and so Eqs. (2.37), (2.38), (2.39) and (2.40) can be summarized by the equations

$$\int \bar{\Phi}_r(x)\Phi_s(x)\,\mathrm{d}x = \delta_{rs} \tag{2.41}$$

$$\iint \bar{\Phi}_r(x)K(x, y)\Phi_s(y)\,\mathrm{d}x\,\mathrm{d}y = \frac{\delta_{rs}}{\lambda_r} \tag{2.42}$$

The eigenfunctions associated with a Hermitian (and hence a symmetric) kernel thus form an orthonormal set.

It must be realized that a kernel need not necessarily have eigenvalues. An example of such a system has been discussed in Example 2.6 and so it is necessary to determine sufficient conditions for them to exist. It may in fact be shown (the proof is rather long, and is given in Appendix E) that an eigenvalue exists if the kernel is Hermitian and is also positive definite, that is

$$\iint \Phi(x)K(x, y)\phi(y)\,\mathrm{d}x\,\mathrm{d}y > 0$$

unless $\phi(x)$ is identically zero.

An eigenvalue will also exist if the kernel is negative definite that is

$$\iint \Phi(x)K(x, y)\phi(y)\,\mathrm{d}x\,\mathrm{d}y < 0$$

unless $\phi(x)$ is identically zero.

In particular, these statements are true for symmetric kernels which are either positive or negative definite.

Suppose that λ_1 and $\Phi_1(x)$ are respectively an eigenvalue and the normalized associated eigenfunction of the system defined by the integral equation

$$\phi(x) = \lambda \int K(x, y)\phi(y)\,\mathrm{d}y$$

Consider now the 'shortened' Hermitian kernel

$$R_1(x, y) = K(x, y) - \frac{\Phi_1(x)\Phi_1(y)}{\lambda_1}$$

If this is not identically zero, it will have at least one eigenvalue λ_2 with an associated eigenfunction $\Phi_2(x)$. Even if λ_2 is the same as λ_1, $\Phi_1(x)$ cannot be an eigenfunction of the shortened kernel R_1, for

$$\int R_1(x, y)\Phi_1(y)\,\mathrm{d}y = \int K(x, y)\Phi_1(y)\,\mathrm{d}y - \frac{\Phi_1(x)}{\lambda_1}\int |\Phi_1(y)|^2\,\mathrm{d}y \tag{2.43}$$

It can be seen that the right hand side of Eq. (2.43) is identically zero.

Two possibilities can arise. The process can terminate after n steps and

$$R_n(x, y) = K(x, y) - \sum_{r=1}^{n} \frac{\Phi_r(x)\bar{\Phi}_r(y)}{\lambda_r}$$

vanishes. In this case the kernel is degenerate, being of the form

$$\begin{aligned} K(x, y) &= \sum_{r=1}^{n} \frac{\Phi_r(x)\bar{\Phi}_r(y)}{\lambda_r} \\ &= \sum_{r=1}^{n} \int K(x, z)\Phi_r(z)\,\mathrm{d}z\,\bar{\Phi}_r(y) \end{aligned}$$

and having a finite number n of eigenvalues and eigenfunctions.

The other possibility is that the process does not terminate and there are an infinite number of eigenvalues and eigenfunctions. When this happens, it is possible to write

$$\begin{aligned} R_m(x) &= \int |R_m(x, y)|^2\,\mathrm{d}x \\ &= \int\Bigg[\bar{K}(x, y)K(x, y) - \bar{K}(x, y)\sum_{r=1}^{m} \frac{\Phi_r(x)\bar{\Phi}_r(y)}{\lambda_r} \\ &\quad - \bar{K}(x, y)\sum_{r=1}^{m} \frac{\Phi_r(x)\bar{\Phi}_r(y)}{\lambda_r} + \sum_{r=1}^{m} \frac{\Phi_r(x)\bar{\Phi}_r(y)}{\lambda_r}\sum_{s=1}^{m} \frac{\Phi_s(x)\bar{\Phi}_s(y)}{\lambda_s}\Bigg]\mathrm{d}y \\ &= [A(x)]^2 - \sum_{r=1}^{m} \frac{\Phi_r(x)\bar{\Phi}_r(x)}{\lambda_r^2} \end{aligned} \tag{2.44}$$

where

$$[A(x)]^2 = \int |K(x, y)|^2 \, dy$$

Now $R_m(x) \geqslant 0$ and so it follows that

$$[A(x)]^2 - \sum_{r=1}^{m} \frac{\Phi_r(x)\bar{\Phi}_r(x)}{\lambda_r^2} \geqslant 0$$

and so the series

$$\sum_{r=1}^{\infty} \frac{\Phi_r(x)\bar{\Phi}_r(x)}{\lambda_r^2}$$

must converge to a sum which is less than or equal to $[A(x)]^2$. Now integrate the relation (2.44)

$$0 \leqslant \int R_n(x) \, dx = \int [A(x)]^2 \, dx - \sum_{r=1}^{m} \frac{1}{\lambda_r^2}$$

It follows immediately that the series

$$\sum_{r=1}^{\infty} \lambda_r^{-2}$$

is convergent and that λ_r tends to infinity.

The question now arises as to whether the series

$$K^*(x, y) = \sum_{r=1}^{\infty} \frac{\Phi_r(x)\bar{\Phi}_r(y)}{\lambda_r} = \sum_{r=1}^{\infty} \Phi_r(x) \int \bar{K}(y, z)\bar{\Phi}_r(z) \, dz \qquad (2.45)$$

represents the kernel $K(x, y)$ in any sense. Two questions arise. The first is: Does the series converge? Secondly, if so, does it converge to $K(x, y)$?

Assume in the first place that the series K^* is uniformly convergent. If this is the case, then it can be proved that $K^*(x, y) = K(x, y)$ everywhere within the domain of interest. The proof is as follows:
Let

$$R(x, y) = K(x, y) - K^*(x, y)$$

Then R is orthogonal to all of the eigenfunctions $\Phi_m(x)$ for

$$\begin{aligned} \int R(x, y)\bar{\Phi}_m(x) \, dx &= \int K(x, y)\bar{\Phi}_m(x) \, dx \\ &\quad - \int \sum_{r=1}^{\infty} \Phi_r(x) \int \bar{K}(y, z)\bar{\Phi}_r(z) \, dz \, \bar{\Phi}_m(x) \, dx \\ &= \int K(x, y)\bar{\Phi}_m(x) \, dx - \int \bar{K}(y, z)\bar{\Phi}_m(z) \, dz = 0 \qquad (2.46) \end{aligned}$$

using the Hermitian property of the kernel. Using this result, it will be shown that $R(x, y)$ vanishes identically. This is done by *reductio ad adsurdum.*

Suppose that R does not vanish identically. It must have at least one eigenvalue λ_R and associated eigenfunction Φ_R.

Then

$$\Phi_R(x) = \lambda_R \int R(x, y)\Phi_R(y)\,dy \tag{2.47}$$

Also, Φ_R will be orthogonal to all the eigenfunctions of K, for

$$\int \Phi_R(x)\bar{\Phi}_m(x)\,dx = \lambda_R \iint \bar{\Phi}_m(x)R(x, y)\,\Phi_R(y)\,dx\,dy = 0 \tag{2.48}$$

by virtue of the result (2.46).

Thus

$$\int K^*(x, y)\Phi_R(y)\,dy = \int \sum_{r=1}^{\infty} \frac{\Phi_r(x)\bar{\Phi}_r(y)}{\lambda_r}\Phi_R(y)\,dy = 0 \tag{2.49}$$

The term by term integration is legitimate by virtue of the uniform convergence of the series (2.45).

Equation (2.47) can be rewritten as

$$\begin{aligned}\Phi_R(x) &= \lambda_R \int [K(x, y) - K^*(x, y)]\Phi_R(y)\,dy \\ &= \lambda_R \int K(x, y)\Phi_R(y)\,dy \end{aligned} \tag{2.50}$$

Thus Φ_R is an eigenfunction of the original kernel K and from Eq. (2.48) it is orthogonal to itself, and so

$$\int \Phi_R(x)\bar{\Phi}_R(x)\,dx = 0$$

This is possible only if $\Phi_R(x)$ is zero almost everywhere. Thus $R(x, y)$ must be zero, that is

$$K(x, y) = \sum_{r=1}^{\infty} \frac{\Phi_r(x)\bar{\Phi}_r(y)}{\lambda_r} \tag{2.51}$$

It will be clear that the above analysis holds even if the set of eigenfunctions $\{\Phi_r\}$ is not complete. Alternatively

$$K(x, y) = \sum_{r=1}^{\infty} \mu_r \Phi_r(x)\bar{\Phi}_r(y)$$

and the case of a separable kernel corresponds to all except a finite number of the μ_r being zero. The above holds even if the series (2.45) is merely convergent in mean to $K^*(x, y)$ (see Appendix F). In this case, the proof of the result (2.49) must take a different form.

Let

$$J = \int K^*(x, y)\Phi_R(y)\,dy = \int\left[K^*(x, y) - \sum_{r=1}^{n} \frac{\Phi_r(x)\bar{\Phi}_r(y)}{\lambda_r}\right]\Phi_R(y)\,dy + \int \sum_{r=1}^{n} \frac{\Phi_r(x)\bar{\Phi}_r(y)}{\lambda_r}\Phi_R(y)\,dy$$

$$= \int\left[K^*(x, y) - \sum_{r=1}^{n} \frac{\Phi_r(x)\bar{\Phi}_r(y)}{\lambda_r}\Phi_R(y)\right]dy$$

by virtue of the result (2.48). This is true for all n. Now, using the Bunyakovskii–Cauchy–Schwarz inequality, it follows that

$$|J|^2 \leqslant \int\left|K^*(x, y) - \sum_{r=1}^{n} \frac{\Phi_r(x)\bar{\Phi}_r(y)}{\lambda_r}\right|^2 dy \int |\Phi_R(y)|^2\,dy$$

and because of the convergence in mean property the first term tends to zero as n becomes large. It follows now that

$$K(x, y) = \lim_{n\to\infty} \sum_{r=1}^{n} \frac{\Phi_r(x)\bar{\Phi}_r(y)}{\lambda_r} \tag{2.52}$$

almost everywhere in the domain of interest. The formula (2.52) is termed the bilinear formula.

Example 2.10

Find the eigenvalues and eigenfunctions associated with the integral equation

$$\phi(x) = \lambda \int_0^{\pi} K(x, y)\Phi(y)\,dy$$

where

$$K(x, y) = \cos x \sin y \qquad 0 \leqslant x \leqslant y \leqslant \pi$$
$$= \cos y \sin x \qquad 0 \leqslant y \leqslant x \leqslant \pi$$

K is clearly symmetric and the Hermitian theory applies. The equation can be rewritten

$$\phi(x) = \lambda \sin x \int_0^x \cos y\,\phi(y)\,dy + \lambda \cos x \int_x^{\pi} \sin y\,\phi(y)\,dy$$

By differentiating twice with respect to x, it follows that

$$\phi''(x)+(1-\lambda)\phi(x) = 0$$

also $\phi(x)$ must satisfy the boundary conditions $\phi'(0) = 0$, $\phi(\pi) = 0$. It can easily be seen that these conditions cannot be satisfied for any $\lambda \geqslant 1$.

The equation $\phi''(x)+\gamma^2\phi(x) = 0$ will satisfy these conditions with a solution of the form $\cos \gamma x$, provided that $\gamma = \mathrm{r}-\frac{1}{2}$, where r is a positive integer. Thus the eigenvalues will be given by

$$1-\lambda_r = (r-\tfrac{1}{2})^2,\ \lambda_r = 1-(r-\tfrac{1}{2})^2$$

and the eigenfunctions will be proportional to $\cos(r-\frac{1}{2})x$.

Now

$$\int_0^\pi [\cos(r-\tfrac{1}{2})x]^2\,\mathrm{d}x = \int_0^\pi \left(\frac{1+\cos(2r-1)x}{2}\right)\mathrm{d}x = \frac{\pi}{2}$$

and so the normalized eigenfunctions will be

$$\sqrt{\left(\frac{2}{\pi}\right)}\cos(r-\tfrac{1}{2})x$$

It follows that

$$K(x,y) = \frac{2}{\pi}\sum_{r=1}^{\infty}\frac{\cos(r-\frac{1}{2})x\cos(r-\frac{1}{2})y}{1-(r-\frac{1}{2})^2}$$

It is easy to see that this series is absolutely convergent.

Example 2.11

Find the eigenvalues and eigenfunctions of the integral equation

$$\phi(x) = \lambda\int_{-\pi}^{\pi}\log\{1-2\alpha\cos(x-y)+\alpha^2\}\phi(y)\,\mathrm{d}y$$

where α is real and $0 < \alpha < 1$.

The kernel is clearly symmetrical:

$$\begin{aligned}\log(1-2\alpha\cos z+\alpha^2) &= \log(1-\alpha\,\mathrm{e}^{iz})+\log(1-\alpha\,\mathrm{e}^{-iz})\\ &= -\sum_{r=1}^{\infty}\frac{\alpha^r\,\mathrm{e}^{irz}}{n}-\sum_{r=1}^{\infty}\frac{\alpha^r\,\mathrm{e}^{-irz}}{r}\\ &= -\sum_{r=1}^{\infty}\frac{2\alpha^r\cos rz}{r}\end{aligned}$$

Thus

$$\log\{1-2\alpha\cos(x-y)+\alpha^2\} = -2\sum_{r=1}^{\infty}\alpha^r\left(\frac{\cos rx\cos ry+\sin rx\sin ry}{r}\right)$$

The eigenfunctions are clearly $\cos rx$ and $\sin rx$. They are not however normalized.

Now

$$\int_{-\pi}^{\pi} \cos^2 rx \, dx = \int_{-\pi}^{\pi} \left(\frac{1-\cos 2rx}{2}\right) = \pi$$

and so the complete orthonormal set is

$$\Phi_r(x) = \sqrt{\left(\frac{1}{\pi}\right)} \cos rx \quad \text{or} \quad \sqrt{\left(\frac{1}{\pi}\right)} \sin rx$$

Thus

$$\log\{1-2\alpha\cos(x-y)+\alpha^2\} = -\sum_{r=1}^{\infty} \frac{2\pi\alpha^r}{r}\frac{1}{\pi}(\cos rx \cos ry + \sin rx \sin ry)$$

Comparing this expansion with the result

$$K(x,y) = \sum_{r=1}^{\infty} \frac{\Phi_r(x)\bar{\Phi}_r(y)}{\lambda_r}$$

it can be seen that the eigenvalues are of the form $-r/(2\pi\alpha^r)$, r a positive integer, that each eigenvalue occurs twice, and that it has corresponding to it the two orthonormal eigenfunctions

$$\left(\frac{1}{\pi}\right)^{\frac{1}{2}} \sin rx, \quad \left(\frac{1}{\pi}\right)^{\frac{1}{2}} \cos rx$$

If $0 < \alpha < 1$, the series is always absolutely convergent.

The result (2.51) can be extended to iterated kernels for

$$\begin{aligned} K_2(x,y) &= \int K(x,z)K(z,y)\,dz \\ &= \int \sum_{r=1}^{\infty} \frac{\Phi_r(x)\bar{\Phi}_r(z)}{\lambda_r} \sum_{s=1}^{\infty} \frac{\Phi_s(z)\bar{\Phi}_s(y)}{\lambda_s}\,dz \\ &= \sum_{r=1}^{\infty}\sum_{s=1}^{\infty} \frac{\Phi_r(x)\bar{\Phi}_s(y)}{\lambda_r\lambda_s}\delta_{rs} \\ &= \sum_{r=1}^{\infty} \frac{\Phi_r(x)\bar{\Phi}_r(y)}{\lambda_r^2} \end{aligned} \tag{2.53}$$

The proof above is legitimate if the series (2.51) is absolutely convergent and can be justified also if the convergence is merely in mean. The further result

$$K_n(x,y) = \sum_{r=1}^{\infty} \frac{\Phi_r(x)\bar{\Phi}_r(y)}{\lambda_r^n} \tag{2.54}$$

can be proved in a similar way. The trace of a kernel is defined by the relation

$$\mathrm{Tr}(K) = \int K(x,x)\,\mathrm{d}x \tag{2.55}$$

It can be seen without any difficulty that

$$\mathrm{Tr}(K_n) = \sum_{r=1}^{\infty} \lambda_r^{-n} \tag{2.56}$$

Example 2.12

Find the sum of the series

$$\sum_{r=1}^{\infty} [1-(r-\tfrac{1}{2})^2]^{-1}$$

From Example 2.10,

$$K(x,y) = \frac{2}{\pi}\sum_{r=1}^{\infty} \frac{\cos(r-\frac{1}{2})x\cos(r-\frac{1}{2})y}{1-(r-\frac{1}{2})^2}$$

$$K(x,x) = \cos x \sin x$$

and

$$\mathrm{Tr}(K) = \sum_{r=1}^{\infty} \lambda_r^{-1} = \sum_{r=1}^{\infty} [1-(r-\tfrac{1}{2})^2]^{-1}$$

$$= \int_0^{\pi} \cos x \sin x\,\mathrm{d}x = 0$$

It must be understood that this theory is only applicable to Hermitian (including symmetric) kernels. If a kernel does not have the required property, the results of the theorems need not hold as can be seen from the following example:

Example 2.13

Find the eigenvalues and eigenfunctions of the kernel

$$K(x,y) = \sum_{n=0}^{\infty} k_n \cos nx \cos(n+1)y \qquad -\pi \leqslant x,\ y \leqslant \pi$$

where the k_n are such that the series is convergent

$$\phi(x) = \lambda \int_{-\pi}^{\pi} \sum_{n=0}^{\infty} k_n \cos nx \cos(n+1)y\phi(y)\,\mathrm{d}y$$

Now any eigenfunction will clearly be of the form

$$\sum_{n=0}^{\infty} \phi_n \cos nx$$

Thus

$$\sum_{n=0}^{\infty} \phi_n \cos nx = \int \sum_{n=0}^{\infty} k_n \cos nx \cos (n+1)y \sum_{s=0}^{\infty} \phi_s \cos sy \, dy$$

and

$$\phi_n = \lambda k_n \int_{-\pi}^{\pi} \cos (n+1)y \sum_{s=0}^{\infty} \phi_s \cos sy \, dy$$
$$= \lambda k_n \pi \phi_{n+1}$$

i.e.

$$\phi_{n+1} = \frac{1}{\lambda k_n \pi} \phi_n \qquad \phi_n = \prod_{s=0}^{n} \frac{1}{\lambda k_s} \phi_0$$

Because the series is uniformly convergent k_n will tend to zero for large n, and it follows that ϕ_n will increase without limit and the eigenfunction does not exist. In the absence of an eigenfunction, the concept of eigenvalue is of course meaningless.

2.4 The Hilbert-Schmidt Theorem

The solution of Fredholm equations of the first and second kind is obtained by using the Hilbert-Schmidt theorem. Before formally stating this, consider the relation between the three functions $f(x)$, $g(x)$, $K(x, y)$ defined by the equation

$$f(x) = \int K(x, y) g(y) \, dy$$

Because integration smooths out irregularities, so to speak, it follows that $f(x)$ will be better behaved than $K(x, y)$ and $g(y)$—for example, if $K(x, y)$ is continuous and $g(y)$ is piecewise continuous, $f(x)$ is continuous or if the functions $f(x) . g(x)$ and $K(x, y)$ are represented by series, then the convergence of the series representing $f(x)$ will be better than that of the series representing $K(x, y)$ and $g(y)$.

The Hilbert-Schmidt theorem is as follows: If $f(x)$ can be written in the form

$$f(x) = \int K(x, y) g(y) \, dy$$

where

$$\int |g(y)|^2 \, dy$$

is finite and the sequence

$$\sum_{r=1}^{\infty} \frac{\Phi_r(x)\bar{\Phi}_r(x)}{\lambda_r}$$

is convergent in mean to $K(x, y)$, then $f(x)$ can be expressed as a convergent series in the orthonormal functions associated with the kernel in the form

$$f(x) = \sum_{r=1}^{\infty} f_r \Phi_r(x) \tag{2.57}$$

Formally

$$f(x) = \sum_{r=1}^{\infty} \frac{\Phi_r(x)\bar{\Phi}_r(y)}{\lambda_r} g(y) \, dy$$

$$= \sum_{r=1}^{\infty} (g_r/\lambda_r)\phi_r(x)$$

where

$$g_r = \int g(y)\bar{\Phi}_r(y) \, dy$$

and so

$$f_r = g_r/\lambda_r \tag{2.58}$$

The λ_r increase with r and so the series with coefficients f_r will be more convergent that the series with coefficients g_r. The proof is as follows:

$$f(x) = \int K(x, y) g(y) \, dy$$

$$= \int \left[K(x, y) - \sum_{r=1}^{n} \frac{\Phi_r(x)\bar{\Phi}_r(y)}{\lambda_r} \right] g(y) \, dy + \int \sum_{r=1}^{n} \frac{\Phi_r(x)\bar{\Phi}_r(y)}{\lambda_r} g(y) \, dy$$

Let

$$r_n(x) = f(x) - \sum_{r=1}^{n} f_r \Phi_r(x) = \int \left[K(x, y) - \sum_{r=1}^{n} \frac{\Phi_r(x)\bar{\Phi}_r(y)}{\lambda_r} \right] g(y) \, dy$$

Then

$$|r_n(x)|^2 \leqslant \int \left| K(x, y) - \sum_{r=1}^{n} \frac{\Phi_r(x)\bar{\Phi}_r(y)}{\lambda_r} \right|^2 \mathrm{d}y \,.\int |g(y)|^2 \,\mathrm{d}y$$

Now because the sequence

$$\sum_{r=1}^{n} \frac{\Phi_r(x)\bar{\Phi}_r(y)}{\lambda_r}$$

is convergent in mean to $K(x, y)$

$$\lim_{n \to \infty} \int \left| K(x, y) - \sum_{r=1}^{n} \frac{\Phi_r(x)\bar{\Phi}_r(y)}{\lambda_r} \right|^2 \mathrm{d}y = 0$$

and it follows that

$$\lim_{n \to \infty} \left[f(x) - \sum_{r=1}^{n} f_r \Phi_r(x) \right] = 0$$

and so the result (2.57) is proved.

It will be noted that

$$\sum_{r=n+1}^{\infty} |f_r \Phi_r(x)|^2 = \sum_{r=n+1}^{\infty} \left| \frac{g_r \Phi_r(x)}{\lambda_r} \right|^2 \leqslant \sum_{r=n+1}^{\infty} |g_r^2| \sum_{r=n+1}^{\infty} \frac{|\Phi_r(x)|^2}{\lambda_r^2}$$

Now

$$\sum_{r=n+1}^{\infty} |g_r^2|$$

will tend to zero for large n (see Appendix G) and if the further condition that

$$\int |K(x, y)|^2 \,\mathrm{d}y = \{A(x)^2\} \leqslant N^2$$

is imposed it follows *a fortiori* that

$$\sum_{r=n+1}^{\infty} \frac{|\phi_r(x)|^2}{\lambda_r^2} \leqslant N^2$$

and that

$$\sum_{r=n+1}^{\infty} |f_r \Phi_r(x)|^2$$

will also tend to zero and the series (2.57) is absolutely convergent.

Consider now the integral equation of the first kind

$$f(x) = \int K(x, y)\phi(y)\,\mathrm{d}y$$

where $f(x)$ is given. The nature of the solution of this integral equation depends on whether the set of eigenfunctions associated with the kernel is complete (Appendix G). Indeed, if the set is incomplete, there may not be a solution.

Consider first the case of when the set is complete. In this case $f(x)$ can be written in the form

$$f(x) = \sum_{r=1}^{\infty} f_r \Phi_r(x) = \sum_{r=1}^{\infty} \int f(z)\bar{\Phi}_r(z)\,\mathrm{d}z\,\Phi_r(x)$$

The kernel is of the form

$$K(x, y) = \sum_{r=1}^{\infty} \frac{\Phi_r(x)\bar{\Phi}_r(y)}{\lambda_r}$$

and the function $\phi(y)$ can be expanded in the form

$$\phi(y) = \sum_{r=1}^{\infty} c_r \Phi_r(y)$$

It follows that

$$f(x) = \int K(x, y)\phi(y)\,\mathrm{d}y = \int \sum_{r=1}^{\infty} \frac{\Phi_r(x)\bar{\Phi}_r(y)}{\lambda_r} \sum_{s=1}^{\infty} c_r \Phi_s(y)\,\mathrm{d}y$$

$$= \sum_{r=1}^{\infty} (c_r/\lambda_r)\phi_r(x)$$

whence

$$c_r = \lambda_r f_r = \lambda_r \int f(z)\bar{\Phi}_r(z)\,\mathrm{d}z$$

and so the solution of the integral equation is given by

$$\phi(x) = \sum_{r=1}^{\infty} \lambda_r \int f(z)\bar{\Phi}_r(z)\,\mathrm{d}z\,\Phi_r(x) \qquad (2.59)$$

and if

$$\int |\phi(x)|^2\,\mathrm{d}x$$

exists; which is the same as saying that

$$\sum_{r=1}^{\infty} \lambda_r^2 f_r^2$$

converges, the conditions of the Hilbert-Schmidt theorem apply and the series (2.59) is the unique solution.

It may be however that the series does not converge even though $f(x)$ is a perfectly well behaved function. This is possible because λ_r increases indefinitely and, although the coefficients f_r are associated with a convergent series, the coefficients $\lambda_r f_r$ may not be. If this happens the solution (2.59) is merely a formal one, and the series solution is associated with a distribution or generalized function (see Appendix I) rather than a function.

Suppose now that the set of eigenfunctions associated with the kernel is not complete, but can be made complete by the addition of a supplementary set of orthonormal eigenfunctions Φ_s^* which can, but need not, be infinite.

The expansion of $f(x)$ is given by

$$\begin{aligned} f(x) &= \sum_{r=1}^{\infty} f_r \Phi_r(x) + \sum_s f_s^* \Phi_s^*(x) \\ &= \sum_{r=1}^{\infty} \int f(z)\bar{\Phi}_r(z)\,\mathrm{d}z\, \Phi_r(x) + \sum_s \int f(z)\bar{\Phi}_s^*(z)\,\mathrm{d}z\, \Phi_s^*(x) \qquad (2.60) \\ K(x,y) &= \sum_{r=1}^{\infty} \frac{\phi_r(x)\bar{\Phi}_r(y)}{\lambda_r} \end{aligned}$$

The sum here has to be over an infinite number of terms. A finite number would correspond to a degenerate kernel. It will be seen in fact that much of the theory when the eigenfunction set associated with the kernel is not complete is similar to that which arises with a degenerate kernel. The solution $\phi(y)$ will be of the form

$$\phi(y) = \sum_{r=1}^{\infty} c_r \Phi_r(y) + \sum_s c_s^* \Phi_s^*(y)$$

Thus

$$\begin{aligned} f(x) &= \int K(x,y)\phi(y)\,\mathrm{d}y \\ &= \int \sum_{r=1}^{\infty} \frac{\Phi_r(x)\bar{\Phi}_r(y)}{\lambda_r} \left[\sum_{r=1}^{\infty} c_r \Phi_r(y) + \sum_s c_s^* \Phi_s^*(y) \right] \mathrm{d}y \\ &= \sum_{r=1}^{\infty} (c_r/\lambda_r)\Phi_r(x) \qquad (2.61) \end{aligned}$$

Comparison of Eqs. (2.60) and (2.61) shows that a solution is possible only if all f_s^* are zero; that is

$$\int f(z)\bar{\Phi}_s^*(z)\,\mathrm{d}z = 0$$

for all s. If this is not the case, the integral equation is not self-consistent, and no solution can exist.

When the solution does exist, it is again given by

$$\phi(y) = \sum_{r=1}^{\infty} \lambda_r \int f(z)\bar{\Phi}_r(z)\,dz\,\Phi_r(y)$$

provided that the series converges. This solution however is not unique, and a complementary function of the form

$$\sum_s c_s^* \Phi_s^*(y)$$

where the c_s^* are arbitrary, must be included. The case where the set of eigenfunctions associated with the kernel is incomplete can be interpreted as follows. If $\mu_r = \lambda_r^{-1}$, the kernel can be written in the form

$$K(x, y) = \sum_{r=1}^{\infty} \mu_r \phi_r(x)\bar{\phi}_r(y) + \sum_s \mu_s^* \Phi_s^*(x)\bar{\Phi}_s^*(y)$$

where all the μ_s^* are zero. (The μ_r tend to zero as r becomes large, but are never actually zero.)

Example 2.14

Find a Fourier series solution for the integral equation

$$f(x) = \frac{1}{2\pi}\int_{-\pi}^{\pi} \frac{1-\alpha^2}{1-2\alpha\cos(x-y)+\alpha^2}\phi(y)\,dy \qquad \begin{array}{c} 0 < \alpha < 1 \\ -\pi \leqslant x \leqslant \pi \end{array}$$

From the results of Exercise 19 at the end of this chapter, it is possible to write

$$\frac{1}{2\pi}\;\frac{1-\alpha^2}{1-2\alpha\cos(x-y)+\alpha^2} = \frac{1}{2\pi} + \frac{1}{\pi}\sum_{n=1}^{\infty} \alpha^n(\cos nx\cos ny + \sin nx \sin ny)$$

and the series is absolutely convergent.

The Fourier series for $f(x)$ is

$$\tfrac{1}{2}a_0 + \sum_{n=1}^{\infty} (a_n \cos nx + b_n \sin nx)$$

where

$$a_n = \frac{1}{\pi}\int_{-\pi}^{\pi} f(z)\cos nz\,dz \qquad n \geqslant 0$$

$$= \frac{1}{\pi}\int_{-\pi}^{\pi} f(z)\sin nz\,dz \qquad n > 0$$

Let

$$\phi(y) = \tfrac{1}{2}c_0 + \sum_{n=1}^{\infty} (c_n \cos ny + d_n \sin ny)$$

Then

$$\int_{-\pi}^{\pi} \left[\tfrac{1}{2}\pi + \frac{1}{\pi} \sum_{n=1}^{\infty} \alpha^n (\cos nx \cos ny + \sin nx \sin ny) \right]$$

$$\times \left[\tfrac{1}{2}c_0 + \sum_{n=1}^{\infty} (c_n \cos ny + d_n \sin ny) \right] \mathrm{d}y = \tfrac{1}{2}c_0 + \sum_{n=1}^{\infty} \alpha^n (c_n \cos nx + d_n \sin nx)$$

and so

$$a_n = c_n \alpha^n, \qquad b_n = d_n \alpha^n$$

Thus the solution of the integral equation is given by the series

$$\tfrac{1}{2}a_0 + \sum_{n=1}^{\infty} \alpha^{-n} (a_n \cos ny + b_n \sin ny)$$

if it is convergent.

Example 2.15

Solve the integral equation

$$\log(1 - 2\beta \cos x + \beta^2) = \frac{1}{\pi} \int_{-\pi}^{\pi} \log\{1 - 2\alpha \cos(x-y) + \alpha^2\} \phi(y) \, \mathrm{d}y$$

$$-\pi \leqslant x \leqslant \pi$$

where α and β are real and positive.

This problem is an example of one where the nature of the solution is dependent upon the values of the parameters α and β. In one case only the solution is obvious. When $\alpha = \beta$, the formal solution is given by

$$\phi(y) = \pi\delta(y)$$

where $\delta(y)$ is the Dirac delta function, and is in fact a distribution and not a function (see Appendix I). This case will be excluded from further consideration.

Now if $\alpha \leqslant 1$,

$$\log(1 - 2\alpha \cos x + \alpha^2) = -\sum_{n=1}^{\infty} \frac{2\alpha^n}{n} \cos nx$$

and if $\alpha > 1$,

$$\log(1 - 2\alpha \cos x + \alpha^2) = 2\log\alpha - \sum_{n=1}^{\infty} \frac{2\alpha^{-n}}{n} \cos nx$$

If $\beta \leqslant 1$,

$$\log\{1-2\beta\cos(x-y)+\beta^2\} = -2\sum_{n=1}^{\infty}\frac{\beta^n}{n}(\cos nx\cos ny+\sin nx\sin ny)$$

In this case, the set of orthonormal functions associated with the kernel is incomplete as there is no constant term.

If $\beta > 1$,

$$\log\{1-2\beta\cos(x-y)+\beta^2\} = 2\log\beta-2\sum_{n=1}^{\infty}\frac{\beta^{-n}}{n}(\cos nx\cos ny+\sin nx\sin ny)$$

Let

$$\phi(y) = \tfrac{1}{2}c_0+\sum_{n=1}^{\infty}(c_n\cos ny+d_n\sin ny)$$

If $\beta \leqslant 1$,

$$\frac{1}{\pi}\int_{-\pi}^{\pi}\log\{1-2\beta\cos(x-y)+\beta^2\}\phi(y)\,\mathrm{d}y = -2\sum_{n=1}^{\infty}\frac{\beta^n}{n}(c_n\cos nx+d_n\sin nx)$$

If $\beta > 1$,

$$\frac{1}{\pi}\int_{-\pi}^{\pi}\log\{1-2\beta\cos(x-y)+\beta^2\}\phi(y)\,\mathrm{d}y$$

$$= 2\log\beta c_0-2\sum_{n=1}^{\infty}\frac{\beta^{-n}}{n}(c_n\cos nx+d_n\sin nx)$$

There are four cases left to consider:

(1) $\alpha \leqslant 1, \quad \beta \leqslant 1$

The solution will be given by

$$c_0 \text{ arbitrary}; \quad c_n = (\alpha/\beta)^n, \quad d_n = 0, \quad n > 0$$

The solution will converge if $\alpha < \beta$, being

$$c_0+\sum_{n=1}^{\infty}(\alpha/\beta)^n\cos nx = c_0+\frac{\alpha\beta\cos x-\alpha^2}{\alpha^2-2\alpha\beta\cos x+\beta^2}$$

If $\alpha > \beta$, there is no function satisfying the integral equation.

(2) $\alpha \leqslant 1, \quad \beta > 1$

The solution will be given by

$$c_0 = 0, \quad c_n = (\alpha\beta)^n, \quad d_n = 0, \quad n > 0$$

and the solution will be

$$\sum_{n=1}^{\infty} (\alpha\beta)^n \cos nx$$

which, if $\alpha\beta < 1$, converges to

$$\frac{\alpha\beta \cos x - \alpha^2\beta^2}{1 - 2\alpha\beta \cos x + \alpha^2\beta^2}$$

If $\alpha\beta > 1$, there is no function satisfying the integral equation.

(3) $$\alpha > 1, \quad \beta \leqslant 1$$

The integral equation can be rewritten as

$$2 \log \alpha - 2 \sum_{n=1}^{\infty} \frac{\alpha^{-n}}{n} \cos nx = -2 \sum_{n=1}^{\infty} \frac{\beta^n}{n} (c_n \cos nx + d_n \sin nx)$$

This equation is not self-consistent because of the constant term on the left hand side. There is therefore no solution.

(4) $$\alpha > 1, \quad \beta > 1$$

The solution will be given by

$$c_0 = \log \alpha/(\log \beta); \quad c_n = (\alpha/\beta)^n, \quad d_n = 0, \quad n > 0$$

giving a solution

$$\frac{\log \alpha}{\log \beta} + \sum_{n=1}^{\infty} \left(\frac{\beta}{\alpha}\right)^n \cos nx = \frac{\log \alpha}{\log \beta} + \frac{\alpha\beta \cos x - \beta^2}{\alpha^2 - 2\alpha\beta \cos x + \beta^2}$$

if $\beta < \alpha$. If $\alpha > \beta$, the series does not converge and there is no function satisfying the integral equation.

It can be seen from this example that attention must always be paid to the ranges of the parameters of an integral equation.

The non-homogeneous equation of the second kind can be solved in an analogous manner. Again it is necessary to consider the question of the completeness of the set of orthonormal functions associated with the kernel.

Consider first the case where the set is complete. Then the equation

$$\phi(x) - \lambda \int K(x, y)\phi(y)\,dy = f(x)$$

can be rewritten as

$$\sum_{r=1}^{\infty} c_r \Phi_r(x) - \lambda \int \sum_{r=1}^{\infty} \frac{\Phi_r(x)\Phi_r(y)}{\lambda_r} \sum_{m=1}^{\infty} c_m \Phi_m(y)\,dy = \sum_{r=1}^{\infty} f_r \Phi_r(x) \quad (2.62)$$

or

$$\sum_{r=1}^{\infty}\left(c_r-\frac{\lambda}{\lambda_r}c_r\right)\Phi_r(x)=\sum_{r=1}^{\infty}f_r\phi_r(x)$$

whence

$$c_r=f_r/(1-\lambda/\lambda_r)$$

and

$$\phi(x)=\sum_{r=1}^{\infty}f_r/(1-\lambda/\lambda_r)\Phi_r(x) \tag{2.63}$$

$$=\sum_{r=1}^{\infty}\frac{\lambda_r f(z)\bar{\Phi}_r(z)\,\mathrm{d}z\,\Phi_r(x)}{\lambda-\lambda_r}$$

$$=f(x)-\lambda\int\sum_{r=1}^{\infty}\frac{\Phi_r(x)\bar{\Phi}_r(y)f(y)\,\mathrm{d}y}{\lambda-\lambda_r} \tag{2.64}$$

Comparing Eqs. (2.35), (2.36) and (2.64), it follows that the resolvent kernel

$$R(x,y:\lambda)=-\sum_{n=1}^{\infty}\lambda^{n-1}K_n(x,y)$$

$$=\sum_{r=1}^{\infty}\frac{\Phi_r(x)\bar{\Phi}_r(y)}{\lambda-\lambda_r}$$

$$=-K(x,y)+\lambda\sum_{r=1}^{\infty}\frac{\Phi_r(x)\bar{\Phi}_r(y)}{\lambda_r(\lambda-\lambda_r)} \tag{2.65}$$

This equation could also be derived by expanding as a power series in λ and using the relation (2.54).

The convergence of the series (2.63) will be the same as that of the series expansion for $f(x)$ because

$$\lim_{n\to\infty}\lambda/\lambda_n=0$$

and so the condition for the applicability of the Hilbert-Schmidt theorem to the expression $\int K(x,y)\phi(y)\,\mathrm{d}y$ will be the same as that for the applicability to the expression $\int K(x,y)f(y)\,\mathrm{d}y$, that is $\int|f(y)|^2\,\mathrm{d}y$ is finite and the infinite series associated with the kernel is convergent in mean to it.

The above analysis holds provided that λ is not equal to an eigenvalue λ_p. If this is the case, it is easy to see that a solution exists only if

$$\int f(z)\bar{\Phi}_p(z)\,\mathrm{d}z=0$$

and the solution is of the form

$$\phi(x) = \sum_{\substack{r=1\\ r\neq p}}^{\infty} \frac{\lambda_r \int f(z)\bar{\Phi}_r(z)\,\mathrm{d}z\,\Phi_r(x)}{\lambda_r - \lambda} + c_p\Phi_p(x) \tag{2.66}$$

where c_p is arbitrary.

If however the orthonormal set associated with the kernel is not complete, it must be completed by the addition of the set Φ_s^*. Exactly as for the case of the integral equation of the first kind.

$$\phi(x) = \sum_{r=1}^{\infty} c_r\Phi_r(x) + \sum_s c_s\Phi_s^*(x)$$

$$f(x) = \sum_{r=1}^{\infty} f_r\phi_r(x) + \sum_s f_s\Phi_s^*(x)$$

Thus

$$\sum_{r=1}^{\infty} c_r\Phi_r(x) + \sum_s c_s^*\Phi_s^*(x) - \lambda \int \sum_{r=1}^{\infty} \frac{\Phi_r(x)\bar{\Phi}_r(y)}{\lambda_r}\left[\sum_{r=1}^{\infty} c_r\Phi_r(y) + \sum_s c_s^*\Phi_s^*(y)\right]\mathrm{d}y$$

$$= \sum_{r=1}^{\infty} f_r\Phi_r(x) + \sum_s f_s^*\Phi_s^*(x)$$

and it follows as before that

$$c_r(1-\lambda/\lambda_r) = f_r, \quad c_s^* = f_s^* \tag{2.67}$$

Remarks may be made in this case also similar to the previous one concerning the application of the Hilbert-Schmidt theorem and the case of λ being an eigenvalue.

Example 2.16

Solve the integral equation

$$\phi(x) = \tfrac{1}{4}(\pi^2 - x^2) + \frac{\lambda}{2\pi}\int_{-\pi}^{\pi} \frac{1-\alpha^2}{1-2\alpha\cos(x-t)+\alpha^2}\phi(t)\,\mathrm{d}t$$

$$-\pi \leqslant x \leqslant \pi, \quad 0 < \alpha < 1$$

Under what condition is the solution possible? (Wales)

It can easily be shown that

$$\tfrac{1}{4}(\pi^2 - x^2) = \tfrac{1}{6}\pi^2 + \sum_{n=1}^{\infty} \frac{(-1)^{n-1}\cos nx}{n^2}$$

Furthermore, it follows from Example 2.14 that,

$$\frac{1}{2\pi}\int_{-\pi}^{\pi}\frac{1-\alpha^2}{1-2\alpha\cos(x-t)+\alpha^2}\phi(t)\,\mathrm{d}t = \tfrac{1}{2}c_0+\sum_{n=1}^{\infty}(c_n\cos nx+d_n\sin nx)\alpha^n$$

if

$$\phi(x) = \tfrac{1}{2}c_0+\sum_{n=1}^{\infty}(c_n\cos nx+d_n\sin nx)$$

Thus

$$\tfrac{1}{2}c_0+\sum_{n=1}^{\infty}(c_n\cos nx+d_n\sin nx)$$

$$=\tfrac{1}{6}\pi^2+\sum_{n=1}^{\infty}\frac{(-1)^{n-1}}{n^2}\cos nx+\lambda\left[\tfrac{1}{2}c_0+\sum_{n=1}^{\infty}\alpha^n(c_n\cos nx+d_n\sin nx)\right]$$

It follows that $c_0 = \frac{1}{3}\pi^2/(1-\lambda)$

$$c_n = \frac{(-1)^{n-1}}{n^2(1-\lambda\alpha^n)},\quad d_n = 0,\quad n>0$$

and so

$$\phi(x) = \tfrac{1}{6}\frac{\pi^2}{1-\lambda}+\sum_{n=1}^{\infty}\frac{(-1)^{n-1}}{n^2(1-\lambda\alpha^n)}\cos nx$$

The solution is valid provided that $\lambda \neq \alpha^{-n}$ where n is $\geqslant 0$. If λ has one of these values, there is no solution.

2.5 Hermitization and Symmetricization of Kernels

In the last section, it became clear that it is possible to do a good deal with a Hermitian (or a symmetric) kernel. The question immediately follows as to whether it is possible to transform an integral equation involving an arbitrary kernel into one involving a Hermitian or a symmetric kernel. This can be done as follows.

Suppose that $K(x, y)$ is an arbitrary kernel. The integral equation of the second kind associated with it is

$$\phi(x)-\lambda\int K(x, y)\phi(y)\,\mathrm{d}y = f(x) \tag{2.68}$$

Multiplying by $\bar{K}(x, z)$ and integrating with respect to x over the domain of definition, it follows that

$$\int\bar{K}(x,z)\phi(x)\,\mathrm{d}x-\lambda\int\bar{K}(x, z)K(x, y)\phi(y)\,\mathrm{d}x\,\mathrm{d}y = \int\bar{K}(x, z)f(x)\,\mathrm{d}x$$

Replacing x by y, and z by x, this last equation can be written as

$$\int [\bar{K}(y,x) - \lambda K_L(x,y)]\phi(y)\,dy = \int \bar{K}(y,x) f(y)\,dy \tag{2.69}$$

where

$$K_L(x,y) = \int \bar{K}(z,x) K(z,y)\,dz \tag{2.70}$$

It can easily be verified that $K_L(x,y)$ is Hermitian. Multiplying Eq. (2.69) by $-\lambda$ and adding the result to Eq. (2.68), it follows that

$$\phi(x) - \lambda \int [K(x,y) + \bar{K}(y,x) - \lambda K_L(x,y)]\phi(y)\,dy = f(x) - \lambda \int \bar{K}(y,x) f(y)\,dy \tag{2.71}$$

The kernel here is Hermitian, but depends on λ. From the point of view of the solution of the inhomogeneous equation of the second kind, this does not matter as λ may be put to unity. No difficulty will arise unless unity is an eigenvalue of

$$\phi(x) - \lambda \int [K(x,y) + \bar{K}(y,x) - \lambda K_L(x,y)]\phi(y)\,dy = 0 \tag{2.72}$$

If this is the case, the problem can be dealt with as previously.

Consider now the associated equation

$$\psi(x) - \lambda \int K(y,x)\psi(y)\,dy = g(x) \tag{2.73}$$

It is possible to obtain, in exactly the same way as previously, another equation with a Hermitian kernel.

$$\psi(x) - \lambda \int [\bar{K}(x,y) + K(y,x) - \lambda K_R(x,y)]\psi(y)\,dy = g(x) - \lambda \int \bar{K}(x,y) g(y)\,dy \tag{2.74}$$

where

$$K_R(x,y) = \int \bar{K}(y,z) K(x,z)\,dz \tag{2.75}$$

If K is real, the same process may be carried through, resulting in the two equations:

$$\phi(x) - \lambda \int [K(x,y) + K(y,x) - \lambda K_L(x,y)]\phi(x) = f(x) - \lambda \int K(y,x) f(y)\,dy \tag{2.76}$$

where

$$K_L(x, y) = \int K(z, x)K(z, y)\,\mathrm{d}z$$

and

$$\psi(x) - \lambda \int [K(x, y) + K(y, x) - K_R(x, y)]\psi(y)\,\mathrm{d}y = g(x) - \lambda \int K(x, y)g(y)\,\mathrm{d}y \tag{2.77}$$

where

$$K_R(x, y) = \int K(x, z)K(y, z)\,\mathrm{d}z$$

It will be seen that K_L and K_R, and hence the kernels associated with Eqs. (2.76) and (2.77) are symmetric.

Now

$$\begin{aligned}\int K_L(x, y)\phi(y)\bar{\phi}(x)\,\mathrm{d}x\,\mathrm{d}y &= \iiint \bar{K}(z, x)K(z, y)\phi(y)\bar{\phi}(x)\,\mathrm{d}x\,\mathrm{d}y\,\mathrm{d}z \\ &= \int \left| \int K(z, y)\phi(y)\,\mathrm{d}y \right|^2 \mathrm{d}z > 0 \end{aligned} \tag{2.78}$$

It follows therefore that the Hermitian kernel K_L (and in exactly the same way K_R) is positive definite. (The same remark will hold for the real case of symmetric kernels. From now on, no special reference will be made to symmetric kernels except where necessary.) This means that the eigenvalues of K_L and K_R are all positive and can be written in the form λ_{Ln}^2 and λ_{Rn}^2 respectively where λ_{Ln} and λ_{Rn} are real non-zero numbers which, for convenience, may be taken as positive. It can in fact be proved that K_L and K_R have the same eigenvalues.

Suppose that $\Phi_L(x)$ is an eigenfunction of K_L associated with the eigenvalue λ_L^2.

Let

$$\begin{aligned}\psi_L(x) &= \lambda_L \int K(x, y)\Phi_L(y)\,\mathrm{d}y \\ &= \lambda_L^3 \int K(x, y) \int K_L(y, u)\Phi_L(u)\,\mathrm{d}u\,\mathrm{d}y \\ &= \lambda_L^3 \iiint K(x, y)\bar{K}(z, y)K(z, u)\Phi_L(u)\,\mathrm{d}u\,\mathrm{d}y\,\mathrm{d}z \\ &= \lambda_L^3 \iint K_R(z, x)K(z, u)\Phi_L(u)\,\mathrm{d}z\,\mathrm{d}u \\ &= \lambda_L^2 \int K_R(z, x)\psi_L(z)\,\mathrm{d}z \end{aligned} \tag{2.79}$$

Thus $\psi_L(x)$ is an eigenfunction of K_R and is associated with an eigenvalue λ_L^2 which is in fact an eigenvalue of K_L.

Similarly, if

$$\phi_R(x) = \lambda_R \int K(y,x)\Psi_R(y)\,\mathrm{d}y$$

where Ψ_R is an eigenfunction of K_R associated with the eigenvalue λ_R^2,

$$\phi_R(x) = \lambda_R^2 \int K_L(x,z)\phi_R(z)\,\mathrm{d}z \tag{2.80}$$

and so ϕ_R is an eigenfunction of K_L and it is associated with an eigenvalue λ_R^2 which is in fact an eigenvalue of K_R. Thus the eigenvalues of K_R and K_L are the same and it is therefore possible to write $\lambda_{Rn}^2 = \lambda_{Ln}^2 = \lambda_n^2$.

Suppose that the eigenfunctions of K_R and K_L corresponding to the eigenvalues λ_n^2 are Ψ_n and Φ_n respectively. Then an orthonormal property exists:

$$\begin{aligned}\int \Phi_m(x)\bar{\Phi}_n(x)\,\mathrm{d}x &= \lambda_m\lambda_n \iiint K(y,x)\Psi_m(y)\bar{K}(z,x)\bar{\Psi}_n(z)\,\mathrm{d}x\,\mathrm{d}y\,\mathrm{d}z \\ &= \lambda_m\lambda_n \iint K_R(z,y)\Psi_m(y)\bar{\Psi}_n(z)\,\mathrm{d}y\,\mathrm{d}z \\ &= \frac{\lambda_n}{\lambda_m}\int \Psi_m(z)\bar{\Psi}_n(z)\,\mathrm{d}z \end{aligned} \tag{2.81}$$

In an exactly similar way, it follows that

$$\int \Psi_m(x)\bar{\Psi}_n(x)\,\mathrm{d}x = \frac{\lambda_n}{\lambda_m}\int \Phi_m(z)\bar{\Phi}_n(z)\,\mathrm{d}z \tag{2.82}$$

Comparison of Eqs. (2.81) and (2.82) shows that if λ_m and λ_n are unequal

$$\int \Phi_m(x)\bar{\Phi}_n(x)\,\mathrm{d}x = \int \Psi_m(x)\bar{\Psi}_n(x)\,\mathrm{d}x = 0$$

It is possible to normalize the Φ_m and the $\bar{\Psi}_n$ so that

$$\int |\Phi_m(x)|^2\,\mathrm{d}x = \int |\Psi_m(x)|^2\,\mathrm{d}x = 1$$

and so the Φ_m and Ψ_m will form orthonormal sets.

Thus

$$K_R(x, y) = \sum_{s=1}^{\infty} \lambda_s^{-2} \Psi_s(x)\bar{\Psi}_s(x) \tag{2.83}$$

$$K_L(x, y) = \sum_{s=1}^{\infty} \lambda_s^{-2} \Phi_s(x)\bar{\Phi}_s(x) \tag{2.84}$$

It can be seen very easily that K_R and K_L can be generated by the kernel

$$\sum_{n=1}^{\infty} \frac{\Phi_n(x)\bar{\Psi}_n(y)}{\lambda_n} \tag{2.85}$$

K_R and K_L may however also be generated by any kernel of the form

$$K(x, y) = \sum_{n=1}^{\infty} \pm \frac{\Phi_n(x)\bar{\Psi}_n(y)}{\lambda_n} \tag{2.86}$$

It should be noted that the λ_n are not eigenvalues of K, and indeed a kernel of this type need not have eigenvalues.

It may be noted that integral equations of the first kind can be solved very easily by the use of K_L for if

$$\int K(x, y)\phi(y)\,\mathrm{d}y = f(x)$$

then

$$\begin{aligned} \int K_L(z, y)\phi(y)\,\mathrm{d}y &= \iint \bar{K}(z, x)K(x, y)\phi(y)\,\mathrm{d}y \\ &= \int \bar{K}(z, x)f(x)\,\mathrm{d}x = g(z) \end{aligned} \tag{2.87}$$

and the methods for the solution of Eq. (2.87) are well known. The solution is given by an expression of the form

$$\phi(y) = \sum_{n=1}^{\infty} \lambda_n^2 \int g(z)\bar{\Phi}_n(z)\,\mathrm{d}z \tag{2.88}$$

in the general case, with suitable modifications as discussed previously in special cases.

The solution for the integral equation of the second kind such as Eq. (2.71) is somewhat more complicated. Suppose that

$$\iint [K(x, y) + \bar{K}(y, x)]\bar{\Phi}_r(x)\Phi_s(y)\,\mathrm{d}x\,\mathrm{d}y = k_{rs}$$

then

$$k_{rs} = \bar{k}_{sr}$$

and

$$K(x,y)+\bar{K}(y,x) = \sum_{r=1}^{\infty}\sum_{s=1}^{\infty} k_{rs}\Phi_r(x)\bar{\Phi}_s(y) \tag{2.89}$$

Also, if

$$\int\left[f(x)-\lambda\int \bar{K}(y,x)f(y)\,\mathrm{d}y\right]\bar{\Phi}_n(x)\,\mathrm{d}x = F_n$$

then

$$f(x)-\lambda\int \bar{K}(y,x)f(y)\,\mathrm{d}y = \sum_{n=1}^{\infty} F_n\Phi_n(x) \tag{2.90}$$

Let

$$\phi(x) = \sum_{n=1}^{\infty} c_n\Phi_n(x) \tag{2.91}$$

Then, using results (2.89), (2.90) and (2.91), it follows that Eq. (2.71) assumes the form

$$\sum_{n=1}^{\infty} c_n\Phi_n(x)-\lambda\int \sum_{r=1}^{\infty}\sum_{s=1}^{\infty} k_{rs}\Phi_r(x)\bar{\Phi}_s(y)\sum_{n=1}^{\infty} c_n\Phi_n(y)\,\mathrm{d}y$$
$$+\lambda^2\int\sum_{s=1}^{\infty}\lambda_s^{-2}\Phi_s(x)\bar{\Phi}_s(y)\sum_{n=1}^{\infty}c_n\Phi_n(y)\,\mathrm{d}y = \sum_{n=1}^{\infty}F_n\phi_n(x)$$

Using the result

$$\Phi_m(x)\bar{\Phi}_n(x)\,\mathrm{d}x = \delta_{mn}$$

it follows in the usual way that

$$c_r-\lambda\sum_{s=1}^{\infty}k_{rs}c_s+(\lambda^2/\lambda_s^2)c_r = F_r \tag{2.92}$$

and the solubility of the system depends on whether the k_{rs} are zero for $r \neq s$.

The discussion above has been made on the basis that the set of eigenfunctions Φ_s is complete. If the set is not complete a modification may be made as previously.

Example 2.17

Show that the kernel

$$K(x,y)=\sum_{n=1}^{\infty}\frac{\sin nx\cos ny}{n+i}\qquad -\pi\leqslant x,\quad y\leqslant\pi$$

does not have any eigenvalues, and find the corresponding K_R and K_L.

If an eigenfunction and an eigenvalue exist, they satisfy the equation

$$\Phi(x)=\lambda\int_{-\pi}^{\pi}\sum_{n=1}^{\infty}\frac{\sin nx\cos ny}{n+i}\Phi(y)\,\mathrm{d}y$$

However, $\Phi(x)$ must be of the form

$$\sum_{n=1}^{\infty}c_n\sin nx$$

But

$$\int_{-\pi}^{\pi}\cos ny\sin my\,\mathrm{d}y=0$$

and so

$$\int_{-\pi}^{\pi}\sum_{n=1}^{\infty}\frac{\sin nx\cos ny}{n+i}\sum_{m=1}^{\infty}c_m\sin my\,\mathrm{d}y=0$$

Thus there is no eigenvalue, and hence no eigenfunction associated with the kernel.

Now

$$\begin{aligned}K_R(x,y)&=\int\overline{K}(y,z)K(x,z)\,\mathrm{d}z\\&=\int_{-\pi}^{\pi}\sum_{n=1}^{\infty}\frac{\sin ny\cos nz}{n-i}\sum_{m=1}^{\infty}\frac{\sin mx\cos mz}{m+i}\mathrm{d}z\\&=\sum_{n=1}^{\infty}\frac{\sin nx\sin ny}{n^2+1}\end{aligned}$$

Similarly

$$\begin{aligned}K_L(x,y)&=\int\overline{K}(z,x)K(z,y)\,\mathrm{d}z\\&=\int_{-\pi}^{\pi}\sum_{n=1}^{\infty}\frac{\sin nz\cos nx}{n-i}\sum_{m=1}^{\infty}\frac{\sin mz\cos ny}{n+i}\mathrm{d}z\\&=\pi\sum_{n=1}^{\infty}\frac{\cos mx\cos my}{n^2+1}\end{aligned}$$

The eigenvalues for K_L and K_R are $(n^2+1)/\pi^2$ and the respective normalized eigenfunctions are $(\sin nx)/\sqrt{\pi}$, $\cos nx/\sqrt{\pi}$.

2.6 Solution of Integral Equations with Green's Function Type Kernels

Consider an integral equation of the inhomogeneous second kind defined by

$$\phi(x)-\lambda\int_a^b K(x,y)\phi(y)\,\mathrm{d}y = f(x) \qquad a\leqslant x\leqslant b \tag{2.93}$$

where

$$\begin{aligned} K(x,y) &= \psi_1(y)\psi_2(x) \qquad a\leqslant y\leqslant x\leqslant b \\ &= \psi_1(x)\psi_2(y) \qquad a\leqslant x\leqslant y\leqslant b \end{aligned}$$

where $\psi_1(x)$ and $\psi_2(x)$ obey the respective boundary conditions

$$\alpha\psi_1(a)+\alpha'\psi_1'(a)=0, \quad \beta\psi_2(b)+\beta'\psi_2'(b)=0$$

$K(x, y)$ regarded as a function of x satisfies the boundary conditions

$$\left[\alpha K(x,y)+\alpha'\frac{\partial K}{\partial x}(x,y)\right]_{x=a} = 0$$

and

$$\left[\beta K(x,y)+\beta'\frac{\partial K}{\partial x}(x,y)\right]_{y=b} = 0 \tag{2.94}$$

and in fact has the form of a Green's function for some second order differential equation when the boundary conditions are (2.94) (see Section 1.3).

Equation (2.93) may be rewritten as

$$\phi(x)-\lambda\psi_2(x)\int_a^x \psi_1(y)\phi(y)\mathrm{d}y-\lambda\psi_1(x)\int_x^b \psi_2(y)\phi(y)\mathrm{d}y = f(x) \tag{2.95}$$

Differentiating Eq. (2.95) with respect to x, it follows that

$$\phi(x)-\lambda\psi_2'(x)\int_a^x \psi_1(x)\phi(y)\mathrm{d}y-\lambda\psi_1'(x)\int_x^b \psi_2(y)\phi(y)\mathrm{d}y = f'(x) \tag{2.96}$$

It can be seen that

$$\begin{aligned} \alpha\phi(a)+\alpha'\phi'(a) &= \alpha f(a)+\alpha' f'(a) \\ \beta\phi(b)+\beta'\phi'(b) &= \beta f(b)+\beta' f'(b) \end{aligned} \tag{2.97}$$

Differentiating Eq. (2.96) with respect to x, it follows that

$$\phi''(x)-\lambda\psi_2''(x)\int_a^x \psi_1(y)\phi(y)\mathrm{d}y-\lambda\psi_1''(x)\int_x^b \psi_2(y)\phi(y)\mathrm{d}y$$
$$-\lambda[\psi_2'(x)\psi_1(x)-\psi_1'(x)\psi_2(x)]\phi(x)=f''(x) \quad (2.98)$$

It is possible to eliminate the quantities

$$\int_a^x \psi_1(y)\phi(y)\mathrm{d}y, \quad \text{and} \quad \int_x^b \psi_2(y)\phi(y)\mathrm{d}y$$

and the following differential equation for $\phi(x)$ is obtained

$$0=\begin{vmatrix} \phi''(x)-\lambda[\psi_2'(x)\psi_1(x)-\psi_1'(x)\psi_2(x)]\phi(x)-f''(x) & \psi_1''(x)\psi_2''(x) \\ \phi'(x)-f'(x) & \psi_1'(x)\psi_2'(x) \\ \phi(x)-f(x) & \psi_1(x)\psi_2(x) \end{vmatrix} \quad (2.99)$$

This is a differential equation of the second order and can be solved uniquely for ϕ by using the boundary conditions (2.97). Note that this is always the case as the coefficient of $\phi''(x)$ which is $\psi_1'(x)\psi_2(x)-\psi_1(x)\psi_2'(x)$ cannot vanish as then ψ_1 would be a multiple of ψ_2.

If the equation is homogeneous, that is $f(x)$ vanishes, the problem simply becomes the eigenvalue problem of finding values of λ for which solutions of the differential equation

$$0=\begin{vmatrix} \phi''(x)-\lambda[\psi_2'(x)\psi_1(x)-\psi_1'(x)\psi_2(x)]\phi(x) & \psi_1''(x)\psi_2''(x) \\ \phi'(x) & \psi_1'(x)\psi_2'(x) \\ \phi(x) & \psi_1(x)\psi_2(x) \end{vmatrix} \quad (2.100)$$

can be found which obey the boundary conditions

$$\alpha\phi(a)+\alpha'\phi'(a)=0, \quad \beta\phi(b)+\beta'\phi'(b)=0 \quad (2.101)$$

The integral equation of the first kind can be solved in a similar manner. Expanding the kernel as previously

$$\psi_2(x)\int_a^x \psi_1(y)\phi(y)\mathrm{d}y+\psi_1(x)\int_x^b \psi_2(y)\phi(y)\mathrm{d}y=f(x) \quad (2.102)$$

Differentiating with respect to x, it follows that

$$\psi_2'(x)\int_a^x \psi_1(y)\phi(y)\mathrm{d}y+\psi_1'(x)\int_x^b \psi_2(y)\phi(y)\mathrm{d}y=f'(x) \quad (2.103)$$

Differentiating again with respect to x,

$$\psi_2''(x)\int_a^x \psi_1(y)\phi(y)\mathrm{d}y+\psi_2''(x)\int_x^b \psi_1(y)\phi(y)\mathrm{d}y$$
$$+[\psi_2'(x)\psi_1(x)-\psi_1'(x)\psi_2(x)]\phi = f'' \quad (2.104)$$

It follows in exactly the same way as before that

$$0=\begin{vmatrix} [\psi_1'(x)\psi_2(x)-\psi_2'(x)\psi_1(x)]\phi(x)+f''(x) & \psi_1''(x)\psi_2''(x) \\ f'(x) & \psi_1'(x)\psi_2'(x) \\ f(x) & \psi_1(x)\psi_2(x) \end{vmatrix} \quad (2.105)$$

However, even though Eq. (2.105) does provide a function $\phi(x)$, the actual integral equation may not have a solution, as it may not be self consistent. That this is the case may be seen by using Eqs. (2.103) and (2.104) which give rise to the two conditions:

$$\alpha f(a)+\alpha' f'(a)=0 \quad (2.106a)$$

$$\beta f(b)+\beta' f'(b)=0 \quad (2.106b)$$

If the relations (2.106) do not hold for $f(x)$, the original integral equation is not self consistent and does not have any solution. This is surprising at first sight, but the reason is as follows.

The solution of the differential equation

$$\frac{\mathrm{d}^2 f}{\mathrm{d}x^2}+q(x)\frac{\mathrm{d}f}{\mathrm{d}x}+r(x)f=\phi(x) \qquad a\leqslant x\leqslant b$$

which satisfies the boundary conditions

$$\alpha f(a)+\alpha' f'(a)=0, \quad \beta f(b)+\beta' f'(b)=0$$

is given by

$$f(x)=\int_a^b K(x,y)\phi(y)\mathrm{d}y$$

where $K(x, y)$ satisfies the same boundary conditions and the differential equation

$$\frac{\mathrm{d}^2 K}{\mathrm{d}x^2}+q(x)\frac{\mathrm{d}K}{\mathrm{d}x}+r(x)K=\delta(x-y)$$

K is in fact the Green's function.

Example 2.18

Let
$$K(x, y) = \sin x \cos y \qquad 0 \leqslant x \leqslant y \leqslant \pi$$
$$= \cos x \sin y \qquad 0 \leqslant y \leqslant x \leqslant \pi$$

Discuss the solution of the integral equations

(i) $\phi(x) = \lambda \int_0^\pi K(x, y)\phi(y)\mathrm{d}y + \mathrm{e}^x$

(ii) $\phi(x) = \lambda \int_0^\pi K(x, y)\phi(y)\mathrm{d}y$

(iii) $\int_0^\pi K(x, y)\phi(y)\mathrm{d}y = x - \frac{x^2}{2\pi}$

(iv) $\int_0^\pi K(x, y)\phi(y)\mathrm{d}y = x$

Now $\int_0^\pi K(x, y)\phi(y)\mathrm{d}y = \sin x \int_x^\pi \cos y\, \phi(y)\mathrm{d}y + \cos x \int_0^x \sin y\, \phi(y)\mathrm{d}y$

In this kernel
$$\psi_1(x) = \sin x, \quad \psi_2(x) = \cos x$$
and the boundary conditions are
$$\psi_1(0) = 0, \quad \psi_2'(\pi) = 0$$

$$\frac{\mathrm{d}}{\mathrm{d}x}\int_0^\pi K(x, y)\phi(y)\mathrm{d}y = \cos x \int_x^\pi \cos y\phi(y)\mathrm{d}y - \sin x \int_0^x \sin y\phi(y)\mathrm{d}y$$

$$\int_0^\pi K(0, y)\phi(y)\mathrm{d}y = 0, \quad \left[\frac{\mathrm{d}}{\mathrm{d}x}\int_0^\pi K(x, y)\phi(y)\mathrm{d}y\right]_{x=\pi} = 0$$

$$\frac{\mathrm{d}^2}{\mathrm{d}x^2}\int_0^\pi K(x, y)\phi(y)\mathrm{d}y = -\phi(x) - \int_0^\pi K(x, y)\phi(y)\mathrm{d}y$$

(i) First of all find the boundary conditions to be imposed on ϕ. It follows immediately that
$$\phi(0) = \lambda \int_0^\pi K(0, y)\phi(y)\mathrm{d}y + 1 = 1$$
and
$$\phi'(\pi) = \lambda \left[\frac{\mathrm{d}}{\mathrm{d}x}\int_0^\pi K(x, y)\phi(y)\mathrm{d}y\right]_{x=\pi} + \mathrm{e}^\pi = \mathrm{e}^\pi$$

It also follows that

$$\phi''(x) = \lambda\left[-\phi(x) - \int_0^\pi K(x, y)\phi(y)\mathrm{d}y\right] + \mathrm{e}^x$$

whence $$\phi''(x) + (\lambda+1)\phi(x) = 2\mathrm{e}^x$$

This can easily be solved using the boundary conditions.

(ii) In this case it is easy to see that, in exactly the same way as in (i),

$$\phi''(x) + (\lambda+1)\phi(x) = 0$$

$$\phi(x) = A\sin\sqrt{(\lambda+1)}x + B\cos\sqrt{(\lambda+1)}x$$

For $\phi(x)$ to vanish for $x = 0$, $B = 0$

$$\phi'(x) = A\sqrt{(\lambda+1)}\cos\sqrt{(\lambda+1)}x$$

and $\phi'(\pi)$ will vanish if

$$\sqrt{(\lambda+1)} = \frac{2n-1}{2}$$

where n is a positive integer.

Thus the eigenvalues are

$$\lambda = 1 + \left(\frac{2n-1}{2}\right)^2$$

and the associated eigenfunctions are

$$\sin(2n-1)x/2$$

(iii) The integral equation can be written as

$$\sin x \int_x^\pi \cos y\phi(y)\mathrm{d}y + \cos x \int_0^x \cos y\phi(y)\mathrm{d}y = x - \frac{x^2}{2\pi}$$

Differentiating

$$\cos x \int_x^\pi \cos y\phi(y)\mathrm{d}y - \sin x \int_0^x \cos y\phi(y)\mathrm{d}y = 1 - \frac{x}{\pi}$$

Putting $x = 0$ and $x = \pi$ respectively in these two equations, the satisfactory result $0 = 0$ appears.

Differentiating a second time it follows that

$$-\phi(x) - \int_0^\pi K(x, y)\phi(y)\mathrm{d}y = -\frac{1}{\pi}$$

whence

$$\phi(x) = \frac{1}{\pi} - x + \frac{x^2}{2\pi}$$

(iv) Putting $x = 0$ in the original integral equation gives the result $0 = 0$. However, differentiating with respect to x gives

$$\cos x \int_x^{\pi} \cos y\phi(y)\mathrm{d}y - \sin x \int_0^x \cos y\phi(y)\mathrm{d}y = 1$$

Putting $x = \pi$, gives the result $0 = 1$. There is thus inconsistency and the integral equation does not have any solution at all which is a function. It will be seen however that $\phi(y) = \delta(\pi - y)$ is a formal solution.

2.7 Miscellaneous

(a) Singular Integral Equations

It was pointed out in Section 1.4(a) that, if the domain of definition of the kernel is infinite, or if the kernel has a singularity within its domain of definition which cannot be transformed away as indicated in Section 1.4(a), the integral equation is said to be singular. When this happens most of the theory of this chapter is inapplicable because

$$\int |K(x, y)|^2 \mathrm{d}x, \quad \int |K(x, y)|^2 \mathrm{d}y \quad \text{and} \quad \iint |K(x, y)|^2 \mathrm{d}x\mathrm{d}y$$

are not all finite, and so the conditions for the convergence theorems governing the legitimacy of the processes are not satisfied. Some of the results in Section 2.2 on degenerate kernels will still be applicable and the ideas of Section 2.5 may be useful, but the Hilbert-Schmidt theorem and its consequences will cease to be valid, and the treatment of singular integral equations is somewhat *ad hoc*.

Example 2.19

Find the eigenvalues and eigenfunctions of the integral equation

$$\phi(x) = \int_{-\infty}^{\infty} \mathrm{e}^{-|x-y|}\phi(y)\,\mathrm{d}y$$

$$\int_{-\infty}^{\infty}\int_{-\infty}^{\infty} |K(x, y)|^2 \mathrm{d}x\mathrm{d}y = \int_{-\infty}^{\infty} \mathrm{d}x \int_{-\infty}^{\infty} \mathrm{e}^{-2|x-y|}\,\mathrm{d}y = \int_{-\infty}^{\infty} \mathrm{d}x$$

is not finite, the equation being singular.

The integral equation can be rewritten as

$$\phi(x) = \lambda\left[\mathrm{e}^{-x}\int_{-\infty}^{x} \mathrm{e}^{y}\phi(y)\mathrm{d}y + \mathrm{e}^{x}\int_{x}^{\infty} \mathrm{e}^{-y}\phi(y)\mathrm{d}y\right]$$

This equation can be dealt with in accordance with the methods indicated in Section 2.5, and it follows that

$$\phi''(x)+(2\lambda-1)\phi(x) = 0$$

Let $\lambda = (1-p^2)/2$. Then the solutions will be of the form

$$\begin{aligned} \phi(x) &= A+Bx & p &= 0 \\ &= Ae^{px}+Be^{-px} & p &\neq 0 \end{aligned}$$

The integrals

$$\int_{-\infty}^{x} e^{y}\phi(y)dy \quad \text{and} \quad \int_{x}^{\infty} e^{-y}\phi(y)dy$$

will both converge when $\phi(y) = 1$ or $\phi(y) = y$, and so $\lambda = \frac{1}{2}$ is a double eigenvalue with associated eigenfunctions 1 and x. It will be noted however that, for $p = 0$,

$$\int_{-\infty}^{\infty} |\phi(x)|^2 dx \text{ is not finite}$$

It will be seen also that

$$\int_{\infty}^{x} e^{y}\phi(y)dy$$

converges when $\phi(y) = e^{ky}$, provided that Re $k < 1$, and that

$$\int_{x}^{\infty} e^{y}\phi(y)dy \quad \text{converges when} \quad \text{Re}\, k > -1$$

Thus $\lambda = (1-p^2)/2$ is a double eigenvalue with associated eigenfunctions e^{px}, e^{-px}, provided that $|\text{Re}\, p| < 1$. If p does not obey this condition, the integral in the integral equation does not converge. It will be seen that in this case again

$$\int_{-\infty}^{\infty} |\phi(x)|^2 dx \quad \text{is not finite}$$

If p is purely imaginary, iq say, $\lambda = (1+q^2)/2$ and the eigenfunctions will be $\sin qx$ and $\cos qx$.

Thus, with this differential equation there is not a discrete sequence of eigenvalues, but a continuous spectrum. This is because the kernel is singular.

Example 2.20

Solve the integral equation

$$\frac{1}{\sqrt{(\pi t)}} \int_{0}^{\infty} e^{-x^2/(4t)}\phi(x)dx = 1$$

Let $x^2 = u$, $t = (4p)^{-1}$.

The integral equation assumes the form

$$\left(\frac{p}{\pi}\right)^{\frac{1}{2}} \int_0^\infty e^{-pu} \frac{\phi(u^{\frac{1}{2}})}{u^{\frac{1}{2}}} du = 1$$

or

$$\int_0^\infty e^{-pu} \frac{\phi(u^{\frac{1}{2}})}{u^{\frac{1}{2}}} du = \left(\frac{\pi}{p}\right)^{\frac{1}{2}}$$

Thus the solution of the original equation is equivalent to asking what is the function $F(u) = u^{-\frac{1}{2}}\phi(u^{\frac{1}{2}})$ whose Laplace transform is $(\pi/p)^{\frac{1}{2}}$. This is in fact $u^{-\frac{1}{2}}$.

Thus

$$u^{-\frac{1}{2}}\phi(u^{\frac{1}{2}}) = F(u) = u^{-\frac{1}{2}}$$

whence

$$\phi(u) = 1$$

(b) Integral Equations in More than One Dimension

It will have been observed that, in nearly all of this chapter, the domains of x and y have not been specified. In general, integrals of the form

$$\int K(x, y)\phi(y)dy$$

have been considered. It should be realized that this integral—and all the associated theories of Fredholm integral equations—need not be interpreted as one dimensional problems only. The x and y may in fact represent a set of variables $(x_1, \ldots x_n)$, $(y_1, \ldots y_n)$, and, although an immediate interpretation of

$$\int K(x, y)\phi(y)\, dy \text{ is } \int_a^b K(x, y)\phi(y)\, dy$$

detailed examination of Sections 2.2, 2.3 and 2.4 will show that the theory will hold also if the interpretation is

$$\int_{a_1}^{b_1} \cdots \int_{a_n}^{b_n} K(x_1, \ldots x_n; \quad y_1, \ldots y_n)\, dy, \ldots dy_n$$

$$a_1 \leqslant x_1 \leqslant b_1; \ldots ; a_n \leqslant x_n \leqslant b_n$$

Example 2.21

Solve the integral equation

$$\int_{-\pi}^{\pi} \int_0^\infty \log\{r^2 - 2rr' \cos(\phi - \phi') + r'^2\} P(r', \phi')r'\, dr'\, d\phi' = f(r, \phi)$$

$$0 \leqslant r, \quad -\pi \leqslant \phi \leqslant \pi$$

This is, apart from a dimensional factor, equivalent to finding the two dimensional electrostatic charge distribution P from the two dimensional electrostatic potential f.

The method of procedure is to expand everything in terms of Fourier series in ϕ and ϕ'.

Let

$$r'P(r',\phi') = \frac{1}{2\pi}P_0 + \frac{1}{\pi}\sum_{n=1}^{\infty}(P_n \cos n\phi' + Q_n \sin n\phi')$$

where the P_n and Q_n are functions of r'

$$f(r,\phi) = \tfrac{1}{2}f_0 + \sum_{n=1}^{\infty}(f_n \cos n\phi + g_n \sin n\phi)$$

where

$$f_n = \frac{1}{\pi}\int_{-\pi}^{\pi} f(r,\phi)\cos n\phi \, \mathrm{d}\phi \qquad n \geqslant 0$$

$$g_n = \frac{1}{\pi}\int_{-\pi}^{\pi} f(r,\phi)\sin n\phi \, \mathrm{d}\phi \qquad n > 0$$

f_n and g_n are functions of r.

Now if $0 \leqslant \alpha \leqslant 1$

$$\log\{1-2\alpha\cos(\phi-\phi')+\alpha^2\} = -2\sum_{n=1}^{\infty}\frac{\alpha^n}{n}\cos n\,(\phi-\phi')$$

$$= -2\sum_{n=1}^{\infty}\frac{\alpha^n}{n}(\cos n\phi \cos n\phi' + \sin n\phi \sin n\phi')$$

and so

$$\log(r^2 - 2rr'\cos\phi-\phi' + r'^2)$$

$$= G_0(r,r') + \sum_{m=1}^{\infty} G_{rn}(r,r')(\cos m\phi \cos m\phi' + \sin m\phi \sin n\phi')$$

where

$$G_0(r,r') = 2\log r \qquad r > r'$$

$$= 2\log r' \qquad r < r'$$

$$\left.\begin{aligned} G_m(r,r') &= -\frac{2\alpha^m}{n}\left(\frac{r'}{r}\right)^m \qquad r > r' \\ &= -\frac{2\alpha^m}{n}\left(\frac{r}{r'}\right)^m \qquad r < r' \end{aligned}\right\} \quad m > 0$$

The integral equation now becomes

$$\int_{-\pi}^{\pi}\int_{0}^{\infty}\sum_{m=0}^{\infty} G_m(r,r')(\cos m\phi\cos m\phi'+\sin m\phi\sin m\phi')$$

$$\left[\frac{1}{2\pi}P_0(r')+\frac{1}{\pi}\sum_{n=1}^{\infty}\{P_n(r')\cos n\phi'+Q_n(r')\sin n\phi'\}\right]\mathrm{d}r'\,\mathrm{d}\phi'$$

$$=\tfrac{1}{2}f_0(r)+\sum_{n=1}^{\infty}\{f_n(r)\cos n\phi+g_n(r)\sin n\phi\}$$

The left hand side becomes

$$\int_0^{\infty} G_0(r,r')P_0(r')\,\mathrm{d}r'+\sum_{n=1}^{\infty}\int_0^{\infty} G_n(r,r')P_n(r')\,\mathrm{d}r'\cos n\phi$$

$$+\sum_{n=1}^{\infty}\int_0^{\infty} G_n(r,r')Q_n(r')\,\mathrm{d}r'\sin n\phi$$

It follows therefore that the solution of the original integral equation is equivalent to the solution of the infinite set of integral equations

$$\int_0^{\infty} G_0(r,r')P_0(r')\,\mathrm{d}r' = \tfrac{1}{2}f_0(r)$$

$$\int_0^{\infty} G_n(r,r')P_n(r')\,\mathrm{d}r' = f_n(r) \qquad n>0$$

$$\int_0^{\infty} G_n(r,r')Q_n(r')\,\mathrm{d}r' = g_n(r) \qquad n>0$$

The first of these equations can be written in the form

$$\log r\int_0^{r} P_0(r')\,\mathrm{d}r'+\int_r^{\infty}\log r'\,P_0(r')\,\mathrm{d}r' = \tfrac{1}{4}f_0(r)$$

Differentiating

$$\frac{1}{r}\int_0^{r} P_0(r')\,\mathrm{d}r' = \tfrac{1}{4}f_0'(r)$$

whence

$$P_0(r)=\tfrac{1}{4}\frac{\mathrm{d}}{\mathrm{d}r}\left(r f_0'(r)\right)$$

Certain conditions must however be imposed on f_0. Substituting back in the original equation, and dropping the factor $\frac{1}{4}$

$$\log r \int_0^r \frac{\mathrm{d}}{\mathrm{d}r'}\left(r'\frac{\mathrm{d}f_0(r')}{\mathrm{d}r'}\right)\mathrm{d}r' + \int_r^\infty \log r' \frac{\mathrm{d}}{\mathrm{d}r'}\left\{r'\frac{\mathrm{d}f_0(r')}{\mathrm{d}r'}\right\}\mathrm{d}r'$$

$$= \log r\Big[[r' f_0'(r')]\Big]_0^r + \Big[[r' \log r' f_0'(r')]\Big]_r^\infty - \int_r^\infty f_0'(r')\,\mathrm{d}r'$$

$$= \log r\Big[[r' f_0'(r')]\Big]_0^r + \Big[[f_0'(r') r' \log r' - f_0(r')]\Big]_r^\infty$$

and it can be seen that the condition for this to be equal to $f_0(r)$ is given by

$$\lim_{r\to 0} r f_0(r) = 0$$

and

$$\lim_{r\to\infty} [r \log r f_0'(r) - f_0(r)] = 0$$

Similarly, the second of the equations can be written in the form

$$-\frac{2\alpha^n}{n}\left[\frac{1}{r^n}\int_0^r r'^n P_n(r')\,\mathrm{d}r' + r^n \int_r^\infty \frac{1}{r'^n} P_n(r')\,\mathrm{d}r'\right] = f_n(r)$$

Absorbing a factor

$$-\frac{n}{2\alpha^n}$$

into $f_n(r)$, this may be written

$$\frac{1}{r^n}\int_0^r r'^n P_n(r')\,\mathrm{d}r' + r^n \int_r^\infty \frac{1}{r'^n} P_n(r')\,\mathrm{d}r' = f_n(r)$$

Differentiating with respect to r

$$-\frac{n}{r^{n+1}}\int_0^r r'^n P_n(r')\,\mathrm{d}r' + n r^{n-1} \int_r^\infty \frac{1}{r'^n} P_n(r')\,\mathrm{d}r' = f_n'(r) = r^{n+1} f_n'(r)$$

Differentiating with respect to r again, it follows that

$$\frac{n(n+1)}{r^{n+2}}\int r'^n P_n(r')\,\mathrm{d}r' + n(n-1) r^{n-2} \int_r^\infty \frac{1}{r'^n} P_n(r')\,\mathrm{d}r'$$

$$-\frac{n}{r}P_n(r) - \frac{n}{r}P_n(r) = f_n''(r)$$

Now from the first two equations,

$$\frac{1}{r^n}\int_0^r r'^n P_n(r')\,\mathrm{d}r' = \frac{1}{2n}\{n f_n(r) - r f_n'(r)\}$$

and

$$r^n \int_r^\infty r'^n P_n(r')\,dr' = \frac{1}{2n}\{nf_n(r) + rf_n'(r)\}$$

It follows that

$$\frac{(n+1)}{2r^2}[nf_n - rf_n'] + \frac{(n-1)}{2r^2}[nf_n + rf_n'] - \frac{2n}{r}P_n(r) = f_n''(r)$$

whence

$$P_n(r) = \frac{1}{2n}\left[rf_n''(r) + f_n'(r) - \frac{n^2}{r}f_n\right]$$

The conditions for the convergence of the integrals are left as an exercise for the reader.

The solution for $Q_n(r')$ follows identically and a solution for

$$P(r', \phi')$$

exists provided that the convergence conditions are satisfied.

(c) Weighted Integral Equations

It is possible to extend the idea of an integral over an interval $a \leqslant x \leqslant b$ in the following way: Suppose that there are n points x_r in the interval $a \leqslant x \leqslant b$ such that $a \leqslant x_1 \ldots x_r < x_{r+1} \ldots \leqslant x_n \leqslant b$ and suppose that there is associated with each of these points a positive number w_r, which may be termed the weighting factor.

Suppose that $f(x)$ is integrable over $a \leqslant x \leqslant b$ and that it is continuous over some interval containing each of the points x_r. Then the following generalization of the concept of integral may be defined

$$\int_a^{\bar{b}} f(x)\,dx = \int_a^b f(x)\,dx + \sum_{r=1}^{n} w_r f(x_r) \tag{2.107}$$

It can be seen that, because all the w_r are positive,

$$\int_a^{\bar{b}} f(x)\,dx \geqslant 0 \quad \text{if} \quad f(x) \geqslant 0 \quad \text{in} \quad a \leqslant x \leqslant b$$

By virtue of this (it is not necessary to go into all the details as the reader can easily fill them in for himself) practically all the Fredholm integral equation theory discussed previously in this chapter is valid when the integral sign is interpreted in the sense of Eq. (2.107). Thus it is possible to solve integral equations where

$$\int K(x, y)\phi(y)\,\mathrm{d}y$$

is interpreted as

$$\int_a^b K(x, y)\phi(y)\,\mathrm{d}y + \sum_{r=1}^{n} w_r K(x, x_r)\phi(x_r)$$

by the use of techniques similar to those discussed previously.

Example 2.22

Solve the integral equation

$$\int_0^{\bar{1}} K(x, y)\phi(y)\,\mathrm{d}y = f(x)$$

where

$$\int_0^{\bar{1}} g(y)\,\mathrm{d}y = \int_0^1 g(y)\,\mathrm{d}y + g(\tfrac{1}{2})$$

and

$$\begin{aligned} K(x, y) &= x \qquad 0 \leqslant x \leqslant y \leqslant 1 \\ &= y \qquad 0 \leqslant y \leqslant x \leqslant 1 \end{aligned}$$

The weighted integral equation thus becomes

(a) $$\int_0^x y\phi(y)\,\mathrm{d}y + x\int_x^1 \phi(y)\,\mathrm{d}y + x\phi(\tfrac{1}{2}) = f(x) \qquad 0 \leqslant x \leqslant \tfrac{1}{2}$$

(b) $$\int_0^x y\phi(y)\,\mathrm{d}y + x\int_x^1 \phi(y)\,\mathrm{d}y + \tfrac{1}{2}\phi(\tfrac{1}{2}) = f(x) \qquad \tfrac{1}{2} \leqslant x \leqslant 1$$

Differentiating with respect to x, it follows that

(c) $$\int_x^1 \phi(y)\,\mathrm{d}y + \phi(\tfrac{1}{2}) = f'(x) \qquad 0 \leqslant x < \tfrac{1}{2}$$

(d) $$\int_x^1 \phi(y)\,\mathrm{d}y = f'(x) \qquad \tfrac{1}{2} < x \leqslant 1$$

Differentiating with respect to x again equations (c) and (d) give the equations

(e) $$\phi(x) = -f''(x) \qquad 0 \leqslant x < \tfrac{1}{2}$$

and

(f) $$\phi(x) = -f''(x) \qquad \tfrac{1}{2} < x \leqslant 1$$

respectively.

Now equations (e) and (f) are apparently the same. However, it can be seen from equations (c) and (d) that putting $x = \frac{1}{2}$ will lead to difficulties as there appears to be an inconsistency. The consistency conditions may be developed in the following manner:

Suppose that

$$\begin{aligned} f(x) &= f_1(x) \qquad 0 \leqslant x < \tfrac{1}{2} \\ &= f_2(x) \qquad \tfrac{1}{2} < x \leqslant 1 \end{aligned}$$

Because $\phi(x)$ is to be defined at $x = \frac{1}{2}$, and to be continuous in some interval containing $x = \frac{1}{2}$, it follows from equations (e) and (f) that $f_1''(\frac{1}{2}) = f_2''(\frac{1}{2})$. Putting $x = \frac{1}{2}$ in equations (c) and (d) a further consistency condition arises, namely that $f_1'(\frac{1}{2}) - f_2'(\frac{1}{2}) = \phi(\frac{1}{2}) = f_1''(\frac{1}{2}) = f_2''(\frac{1}{2})$. From equation (d) it follows that $f'(1) = 0$. Putting $x = \frac{1}{2}$ in equations (a) and (b) it follows that $f_1(\frac{1}{2}) = f_2(\frac{1}{2})$, and putting $x = 0$ in equation (a) it follows that $f(0) = 0$. Thus the solution of the integral equation is given by

$$\begin{aligned} \phi(x) &= -f_1''(x) \qquad 0 \leqslant x \leqslant \tfrac{1}{2} \\ &= -f_2''(x) \qquad \tfrac{1}{2} \leqslant x \leqslant 1 \end{aligned}$$

provided that all the consistency conditions are satisfied.

EXERCISES

1. Solve the integral equation

$$\phi(x) = \lambda \int_{-\pi}^{\pi} \sin(x+t)\phi(t) + s \qquad -\pi \leqslant s \leqslant \pi$$

for $\phi(s)$ and comment on the solution when $\lambda = \pm 1/\pi$. (Wales)

2. Solve the integral equation

$$\phi(x) = \cos \alpha x + \int_{-\pi}^{\pi} \cos(x+2y)\phi(y)\,dy \qquad -\pi \leqslant x \leqslant \pi$$

(Wales)

3. Find the eigenvalues and orthonormal eigenfunctions of the integral equation

$$\phi(x) = \lambda \int_{-\pi}^{\pi} (\cos x \cos 5y + \cos 5x \cos y)\phi(y)\,dy \qquad -\pi \leqslant x \leqslant \pi$$

(Wales)

4. Show that the solution of the integral equation

$$\phi(x) - \lambda \int_0^1 x\,e^t\,\phi(t) = f(x) \qquad \lambda \neq 1$$

is

$$\phi(x) = f(x) + \frac{\lambda}{1-\lambda} \int_0^1 x\,e^t f(t)\,dt$$

and comment on the case $\lambda = 1$.

5. Show that the integral equation

$$\phi(s) = \lambda \int_0^\pi (\sin s \sin 2t + \sin 3s \sin 4t)\phi(t)\,dt$$

does not have a non-trivial solution.

6. Defining the sequence of functions

$$f_n(x) \text{ by } f_1(x) = f(x), \quad f_{n+1}(x) = f(x) + \lambda \int K(x, y) f_n(y)\,dy, \quad n > 1$$

prove that if

$$\phi(x) - \lambda \int K(x, y)\phi(y)\,dy = f(x)$$

then

$$\phi(x) - \lambda^n \int K_n(x, y)\phi(y)\,dy = f_n(x)$$

7. $K(x, y)$ is a kernel, which has an eigenvalue associated with p eigenfunctions. Prove that

$$p \leqslant |\lambda|^2 \iint |K(x, y)|^2\,dx\,dy$$

8. Prove that the equation

$$\phi(x) = \lambda \int_{-1}^{+1} (x^4 - x^2)(4t^3 + 3t)\phi(t)\,dt$$

does not have any eigenvalues.

9. Show that finding the first eigenvalue of the system defined by the integral equation

$$\phi(x) = \lambda \int K(x, y)\phi(y)\,dy$$

where K is a positive definite Hermitian kernel, is equivalent to finding the maximum value of

$$\iint \bar{\phi}(x)K(x,y)\phi(y)\,dx\,dy$$

when

$$\iint |\phi(x)|^2\,dx = 1$$

10. Suppose that $K(x)$ is an even periodic function of x with period 2π. Show that the eigenvalues and associated eigenfunctions of the kernel of the integral equation

$$\phi(x) = \lambda \int_{-\pi}^{\pi} K(x-y)\phi(y)\,dy$$

are

$$\left[\int_{-\pi}^{\pi} K(x)\cos nx\,dx\right]^{-1}$$

and

$\cos nx$, $\sin nx$ respectively (n being a positive integer)

and

$$\left[\int_{-\pi}^{\pi} K(x)\,dx\right]^{-1}$$

and 1 when n is zero.

11. Let

$$\begin{aligned} K(x,y) &= (1-x)y \qquad 0 \leqslant y \leqslant x \leqslant 1 \\ &= (1-y)x \qquad 0 \leqslant x \leqslant y \leqslant 1 \end{aligned}$$

Prove that

$$K(x,y) = 2\sum_{n=1}^{\infty} \frac{\sin n\pi x \sin n\pi y}{n^2\pi^2}$$

and hence that

$$\sum_{n=1}^{\infty} \frac{1}{n^2} = \frac{\pi^2}{6}$$

12. Prove that an eigenfunction of the integral equation

$$\phi(x) = \lambda \int K(x,y)\phi(y)\,dy$$

where K is bounded, is continuous if it is bounded.

13. Prove that, for a skew-symmetric kernel, $\{K(x, y)+K(y, x) = 0\}$, all the eigenvalues are purely imaginary.

14. Show if λ_r, λ_s are eigenvalues of the kernel $K (r \neq s)$, and if

$$\Phi_r(x) = \lambda_r \int K(x, y)\Phi_r(y)\,\mathrm{d}y$$

$$\Psi_s(x) = \lambda_s \int K(y, x)\Psi_s(y)\,\mathrm{d}y$$

that

$$\int \Phi_r(x)\Psi_s(x)\,\mathrm{d}x = 0$$

$$\iint \Phi_r(x)K(y, x)\Psi_s(y)\,\mathrm{d}x\,\mathrm{d}y = 0$$

15. Prove that the integral equation

$$\phi(x) = \lambda \int K(x, y)p(y)\phi(y)\,\mathrm{d}y + f(x)$$

where $p(y) > 0$ in the domain of interest can be reduced by the transformation

$$\psi(x) = \{p(x)\}^{\frac{1}{2}}\phi(x)$$

to a problem involving a symmetric kernel.

16. Show that for a positive definite kernel, all the eigenvalues are positive.

17. Show that, if $|\alpha| < 1$ the normalized eigenfunctions of the integral equation

$$\phi(x) = \frac{\lambda}{2\pi}\int_{-\pi}^{\pi} \frac{1-\alpha^2}{1-2\alpha\cos(x-t)t\alpha^2}\phi(t)\,\mathrm{d}t \qquad -\pi \leqslant \phi \leqslant \pi$$

are given by

$$\phi_0(x) = \frac{1}{\sqrt{(2\pi)}}$$

$$\phi_{2r-1}(x) = \frac{\sin rx}{\sqrt{\pi}}, \quad \phi_{2r}(x) = \frac{\cos rx}{\sqrt{\pi}} \qquad r \geqslant 1$$

with corresponding eigenvalues

$$\lambda_0 = 1, \quad \lambda_{2r-1} = \frac{1}{\alpha^r}, \quad \lambda_{2r} = \frac{1}{\alpha^r}$$

(Wales)

18. Find the eigenvalues and orthonormal eigenfunctions of the integral equation

$$\phi(x) = \lambda \int_{-\pi}^{\pi} B(x, y)\phi(y)\,dy \qquad -\pi \leqslant x \leqslant \pi$$

where

$$B(x, y) = \sum_{n=1}^{\infty} b_n \sin nx \sin ny$$

Indicate what condition the b_n must satisfy if they are all positive.

(Wales)

19. Solve the integral equation

$$\phi(x) - \lambda \int_{0}^{\pi} K(x, t)\phi(t)\,dt = 3 \qquad 0 \leqslant x \leqslant \pi$$

where

$$\begin{aligned} K(x, t) &= \sin x \cos t \qquad 0 \leqslant x \leqslant t \leqslant \pi \\ &= \sin t \cos x \qquad 0 \leqslant t \leqslant x \leqslant \pi \end{aligned}$$

20. Find the differential equation and boundary conditions associated with the eigenfunctions of the integral equation

$$\phi(x) = \lambda \int_{0}^{1} K(x, y)\phi(y)\,dy \qquad 0 \leqslant x \leqslant 1$$

$$\begin{aligned} K(x, y) &= (x+1)(y-2) \qquad 0 \leqslant x \leqslant y \leqslant 1 \\ &= (y+1)(x-2) \qquad 0 \leqslant y \leqslant x \leqslant 1 \end{aligned}$$

21. Let $\Phi_n(x)$, $\Psi_n(x)$ be sets of orthonormal functions such that

$$\int \Phi_m(x)\Psi_n(x)\,dx = \delta_{mn}$$

Let

$$K(x, y) = \sum_n \frac{\Phi_n(x)\Phi_n(y)}{\lambda_n}$$

Let

$$f_n = \int f(x)\Psi_n(x)\,dx$$

Prove that

(i) the eigenvalues and associated eigenfunctions of the integral equation

$$\phi(x) = \lambda K(x, y)\phi(y)\,dy$$

are λ_n and $\Phi_n(x)$ respectively,

(ii) the solution of the integral equation

$$\int K(x, y)\phi(y)\,dy = f(x)$$

is

$$\phi(y) = \sum_n \lambda_n f_n \Phi_n(y)$$

(iii) the solution of the integral equation

$$\phi(x) = \lambda \int K(x, y)\phi(y)\,dy + f(x)$$

is

$$\phi(x) = \sum_n \frac{f_n \lambda_n}{\lambda_n - \lambda} \Phi_n(x)$$

(Conditions and slight modifications in special cases similar to those of Section 2.3 are applicable.)

22. Solve the integral equations

(i) $$\phi(x) - \int_0^a K(x, y)\phi(y)\,dy = e^{2x}$$

(ii) $$\phi(x) = \lambda \int_0^a K(x, y)\phi(y)\,dy$$

(iii) $$\int_0^a K(x, y)\phi(y)\,dy = x(a-x)$$

where

$$\begin{aligned} K(x, y) &= \sinh(x-a)\sinh y \qquad 0 \leqslant y \leqslant x \leqslant a \\ &= \sinh x \sinh(y-a) \qquad 0 \leqslant x \leqslant y \leqslant a \end{aligned}$$

Comment on the possibility of solving the integral equations

(a) $$\int_0^a K(x, y)\phi(y)\,dy = f(x)$$

(b) $$\int_0^a K(x, y)\phi(y)\,dy = g(x)[g(x)-g(a)]$$

23. Prove that the solution, if it exists, of the integral equation

$$\int_{-\infty}^{\infty} e^{-|x-y|}\phi(y)\,dy = f(x)$$

is

$$\tfrac{1}{2}\{f(x)-f''(x)\}$$

and state the conditions which must be imposed on $f(x)$ for the solution to be valid.

24. Find the conditions that must be imposed on the coefficients k_{mn} in the kernel

$$K(x, y) = \sum_{m=1}^{\infty}\sum_{n=1}^{\infty} k_{mn}\Phi_m(x)\bar{\Phi}_n(y)$$

when the associated iterated kernel $K_R(x, y)$ is given by

$$K_R(x, y) = \sum_{s=1}^{\infty} \lambda_s^{-2}\Phi_s(x)\bar{\Phi}_s(y)$$

25. Prove that the eigenfunctions of the integral equation

$$\phi(s) = \lambda(1-s)\int_0^s \phi(t)\,dt + \lambda\int_s^1 (1-t)\phi(t)\,dt + \lambda(1-s)\phi(0)$$

are of the form $\cos\lambda^{\frac{1}{2}}\sin\lambda^{\frac{1}{2}}x - \sin\lambda^{\frac{1}{2}}\cos\lambda^{\frac{1}{2}}x$ and find what are the possible values of λ.

26. Solve the integral equation

$$\frac{1}{\sqrt{(\pi t)}}\int_0^{\infty} e^{-x^2/(4t)}\phi(x)\,dx = t$$

27. Prove that a formal solution to the integral equation

$$\phi(x) = f(x) + \lambda\int K(x, y)\phi(y)\,dy$$

is given by

$$\phi(x) = f(x) + \frac{\lambda\int D(x, y;\lambda)\phi(y)\,dy}{D_0(\lambda)}$$

where

$$D(x, y; \lambda) = \sum_{n=0}^{\infty} \frac{(-\lambda)^n}{n!} G_n(x, y)$$

$$G_n(x, y) = d_n K(x, y) - n \int G_{n-1}(x, z) K(z, y) \, \mathrm{d}z \qquad n \geqslant 0$$

$$D_0(\lambda) = \sum_{n=0}^{\infty} \frac{(-\lambda)^n}{n!} d_n, \quad d_0 = 1$$

3 Volterra Integral Equations

3.1 Types of Volterra Equations

A kernel $K(x, y)$ is said to be of Volterra type if

$$K(x,y) = 0, \qquad y > x \tag{3.1}$$

It is clear that such a kernel cannot be symmetric or Hermitian. It will be assumed in this treatment that the kernel is real, but it will be seen, generally speaking, that the theory will hold even when the kernel is complex.

Consider now the three possible types of linear integral equation. They are respectively
the equation of the first type:

$$f(x) = \int_0^x K(x,y)\phi(y)\,\mathrm{d}y \tag{3.2}$$

the equation of the seond type:

$$\phi(x) = \lambda \int_0^x K(x,y)\phi(y)\,\mathrm{d}y + f(x) \tag{3.3}$$

and the homogeneous equation of the second type:

$$\phi(x) = \lambda \int_0^x K(x,y)\phi(y)\,\mathrm{d}y \tag{3.4}$$

Rather surprisingly the treatment of Volterra integral equations is somewhat different to that of Fredholm integral equations. The lower limit of integration has been taken as zero, but this is only a matter of convenience and any other limit could be taken with suitable changes being made in the analysis.

The following properties arise immediately:

(i) It is necessary for consistency in the equation of the first type that $f(0) = 0$. Inconsistency may, however, sometimes be resolved by the use of generalized functions.

(ii) Any solution to the equation of the second kind which is obtained, cannot be correct unless $\phi(0) = f(0)$.
(iii) It has been shown previously in Section 1.4(b) that, if K is non-singular, there are not any eigenvalues and eigenfunctions associated with the homogeneous equation (3.4) of the second kind.
(iv) It has been shown previously in Section 1.4(b) that the equation of the first type (3.2) can be differentiated to give the equivalent equation

$$K(x,x)\phi(x)+\int_0^x \frac{\partial K(x,y)}{\partial x}\phi(y)\,\mathrm{d}y = f'(x) \tag{3.5}$$

If $K(x,x)$ is identically zero, there is another equation of the first kind with kernel

$$\frac{\partial K}{\partial x}(x,y)$$

If $K(x,x)$ is not identically zero, Eq. (3.5) can be rewritten in the form

$$\phi(x) = \int_0^x K^*(x,y)\phi(y)\,\mathrm{d}y+g(x) \tag{3.6}$$

where

$$K^*(x,y) = -\frac{\partial K(x,y)}{\partial x}\bigg/ K(x,x)$$

and

$$g(x) = f'(x)/K(x,x)$$

In this case

$$\phi(0) = g(0)$$

If $K(x,x)$ is zero for some x, then $K^*(x,y)$ is singular.

The search for the solution of Volterra equations often proceeds in a more intuitive manner than that for the solution of Fredholm equations.

Example 3.1

Solve the integral equation

$$x^2 = \int_0^x \sin a(x-y)\phi(y)\,\mathrm{d}y \qquad \text{(i)} \qquad a \neq 0$$

Differentiating with respect to x, it follows that

$$2x = a\int_0^x \cos a(x-y)\phi(y)\,\mathrm{d}y \qquad \text{(ii)}$$

and after differentiating with respect to x a second time, the result follows

$$2 = a\phi(x) - a^2 \int_0^x \sin a(x-y)\phi(y)\,dy = a\phi(x) - a^2x^2$$

whence

$$\phi(x) = a^{-1}(2 + a^2x^2)$$

It is in fact luck that this equation can be solved because the kernel, when differentiated twice, gives a multiple of itself. It will be observed that the consistency conditions for $x = 0$ are obeyed in Eqs. (i) and (ii).

Example 3.2

Solve the integral equation

$$1 = \int_0^x \cos \alpha(x-y)\phi(y)\,dy$$

Putting $x = 0$, the two sides of the equation are inconsistent. It may however be verified that it is satisfied formally by

$$\phi(x) = \delta(x) + \alpha^2 x$$

which is a solution in terms of generalized functions.

Example 3.3

Solve the integral equation

$$\phi(x) = 3\int_0^x \cos(x-y)\phi(y)\,dy + e^x$$

It can be seen that $\phi(0) = 1$. Differentiating with respect to x, it follows that

$$\phi'(x) = 3\phi(x) - 3\int_0^x \sin(x-y)\phi(y)\,dy + e^x$$

Thus

$$\phi'(0) = 3\phi(0) + 1 = 4$$

Differentiating with respect to x again,

$$\begin{aligned}\phi''(x) &= 3\phi'(x) - 3\int_0^x \cos(x-y)\phi(y)\,dy + e^x \\ &= 3\phi'(x) - \phi(x) + 2e^x\end{aligned}$$

This differential equation can easily be solved.

Consider now an integral equation of the form

$$\phi(x)+\int_0^x \sum_{p=0}^{n-1} k_p(x)u^{p-1}\phi(u)\,du = g(x)$$

This may be rewritten in the form

$$\phi(x)+\sum_{q=1}^{n} l_q(x)\int_0^x \frac{(x-u)^{q-1}}{(q-1)!}\phi(u)\,du = g(x) \tag{3.7}$$

Differentiating with respect to x and using the result of Appendix A,

$$\phi'(x)+l_1(x)\phi(x)+l_1'(x)\int_0^x \phi(u)\,du+\sum_{q=2}^{n} l_q(x)\int_0^x \frac{(x-u)^{q-2}}{(q-2)!}\phi(u)\,du$$
$$+\sum_{q=2}^{n} l_q'(x)\int_0^x \frac{(x-u)^{q-1}}{(q-1)!}\phi(u)\,du = g'(x) \tag{3.8}$$

It is possible to eliminate the quantity

$$\int_0^x (x-u)^{q-1}\phi(u)\,du$$

between Eqs. (3.7) and (3.8) to obtain an equation of the form

$$\phi'(x)+m_1(x)\phi(x)+\sum_{q=2}^{n} m_q(x)\int_0^x \frac{(x-u)^{q-2}}{(q-2)!}\phi(u)\,du = g_1(x)$$

This process may be repeated n times and eventually a differential equation of the form

$$\phi^{(n)}(x)+\sum_{s=1}^{n} a_s(x)\phi^{(n-s)}(x) = g_n(x)$$

is arrived at. The solution of this is associated with the initial conditions

$$\phi(0) = g(0), \quad \phi'(0)+m_1(0)\phi(0) = g_1(0), \quad \text{etc.}$$

Example 3.4

Solve the integral equation

$$\phi(x) = x+1+\int_0^x [1+2(x-y)]\phi(y)\,dy$$

Differentiating once it follows that

$$\phi'(x) = 1+\phi(x)+2\int_0^x \phi(y)\,dy$$

and so

$$\phi(0) = 1, \quad \phi'(0) = 1+\phi(0) = 2$$

Differentiating again it follows that $\phi''(x) = \phi'(x)+2\phi(x)$, and the solution follows.

3.2 Resolvent Kernel of Volterra Equation

Consider the Volterra equation of the second kind

$$\phi(x) = \lambda \int_0^x K(x, y)\phi(y)\,dy + f(x)$$

It is possible to find a solution of this equation as a power series in λ in exactly the same manner as that discussed at the end of Section 2.2. In this case, the iterated kernels are defined by

$$K_n(x, y) = \int_y^x K(x, z)K_{n-1}(z, y)\,dz \qquad n \geqslant 2 \tag{3.9}$$

Define an iterative sequence $\{\phi_n(x)\}$ of functions by the relations

$$\phi_0(x) = f(x)$$

$$\phi_n(x) = \lambda \int_0^x K(x, y)\phi_{n-1}(y)\,dy + f(x)$$

It follows that

$$\phi_n(x) = f(x) + \sum_{s=1}^{n} \lambda^s \int_0^x K_s(x, y)f(y)\,dy \tag{3.10}$$

The sequence $\phi_n(x)$ thus generates a power series in λ

$$\phi(x) = f(x) + \sum_{s=1}^{\infty} \lambda^s \int_0^x K_s(x, y)f(y)\,dy \tag{3.11}$$

$$= f(x) - \lambda \int_0^x R(x, y; \lambda)f(y)\,dy \tag{3.12}$$

and so the resolvent kernel is

$$R(x, y; \lambda) = -\sum_{s=1}^{\infty} \lambda^{s-1} K_s(x, y) \tag{3.13}$$

The problem remains of determining the conditions under which the power series on the right hand side of Eq. (3.11) is convergent.

Suppose that over $0 \leqslant x, y \leqslant l, |K(x,y)| \leqslant k$.
Then

$$|K_2(x,y)| = \left|\int_y^x K(x,z)K(z,y)\,dz\right|$$
$$\leqslant k^2(x-y) \qquad x \geqslant y$$

also

$$K_2(x,y) = 0 \qquad y \leqslant x$$

Similarly

$$|K_3(x,y)| = \left|\int_y^x K_2(x,z)K(z,y)\,dz\right|$$
$$\leqslant k^3 \int_y^x (x-z)\,dz$$
$$= \tfrac{1}{2}k^3(x-y)^2 \qquad x \geqslant y$$

and

$$K_3(x,y) = 0 \qquad x \leqslant y$$

Proceeding in this way it follows that

$$|K_n(x,y)| \leqslant \frac{1}{(n-1)!}k^n(x-y)^{n-1} \qquad x \geqslant y$$
$$= 0 \qquad x \leqslant y$$

Thus the series with the nth term $\lambda^n K_n(x,y)$ is dominated by the series with nth term

$$\frac{\lambda^n}{(n-1)!}k^n(x-y)^{n-1}$$

(An infinite series with positive terms dominates another if each of its terms is greater than the corresponding term of the second series. Thus, if the dominating series converges, so also does the dominated series.)
Now

$$|x-y| \leqslant 2l$$

and so this latter series is dominated by the series whose nth term is

$$\lambda^n k \frac{(2lk)^{n-1}}{(n-1)!}$$

This is the typical term of an exponential series and so it follows that the series (3.12) for $R(x, y: \lambda)$ always converges.

It will be remembered that, for a Fredholm kernel, the power series in λ defining the resolvent kernel was only convergent up to the modulus of the first eigenvalue. Put in another way, the modulus of the first eigenvalue is greater than the moduli of any values of λ for which the series converges. A Volterra kernel is a special case of a Fredholm kernel, for which the resolvent kernel series converges for all values of λ if the original kernel is bounded. This is equivalent to saying that there is not any eigenvalue. The same remarks hold when the kernel is only weakly singular, in which case the iterated kernels eventually become non-singular and a suitable modification can be made to the proof. The uniqueness of the solution (3.12) follows easily because, if $\phi^A(x)$, $\phi^B(x)$ are both solutions, then

$$\phi^A(x) - \phi^B(x) = \lambda \int_0^x K(x, y)[\phi^A(y) - \phi^B(y)]\, dy$$

Because there are no eigenvalues, $\phi^A(x) = \phi^B(x)$.

Example 3.5

Solve the Volterra equation

$$\phi(x) = \lambda \int_0^x e^{k(x-y)} \phi(y)\, dy + f(x)$$

$$K_1(x, y) = e^{k(x-y)}$$

$$K_2(x, y) = \int_y^x e^{k(x-z)} e^{k(z-x)}\, dz = (x-y) e^{k(x-y)}$$

Similarly

$$K_n(x, y) = e^{k(x-y)} \frac{(x-y)^{n-1}}{(n-1)!}$$

The resolvent kernel is thus

$$-\sum_{n=1}^{\infty} \lambda^{n-1} K_n(x, y) = -e^{k(x-y)} e^{\lambda(x-y)}$$

and so

$$\phi(x) = f(x) + \lambda \int_0^x e^{(k+\lambda)(x-y)} f(y)\, dy$$

Example 3.6

Find the resolvent kernel of the integral equation

$$\phi(x) = f(x) + \int_0^x (x-t)\phi(t)\,dt$$

This resolvent kernel will be found in an indirect manner. (λ has been assumed unity).

Differentiating with respect to x

$$\phi'(x) = f'(x) + \int_0^x \phi(t)\,dt$$

$$\phi''(x) = f''(x) + \phi(x)$$

where

$$\phi(0) = f(0), \qquad \phi'(0) = f'(0)$$

It may be verified that the solution of the differential equation under the given initial conditions is

$$\phi(x) = f(x) - \int_0^x \sinh(x-y)\phi(y)\,dy$$

and it follows that the resolvent kernel is $\sinh(x-y)$.

It may be remarked that similar considerations apply to Volterra equations in more than one dimension. For example, the inhomogeneous integral equation of the second kind

$$\phi(x_1, \ldots, x_n) = f(x_1, \ldots, x_n) + \lambda \int_0^{x_1} \cdots \int_0^{x_n} K(x_1, \ldots, x_n; y_1, \ldots, y_n)\phi(y_1, \ldots, y_n)\,dy_1, \ldots dy_n \tag{3.14}$$

may be solved by means of an iterative sequence

$$\phi_0(x) = f(x)$$

$$\phi_m(x) = f(x) + \lambda \int_0^x K(x, y)\phi_{m-1}(y)\,dy \tag{3.15}$$

$$x = (x_1, \ldots, x_n), \quad y = (y_1, \ldots, y_n)$$

giving, in exactly the same way as before when K is non-singular, a unique solution of the form

$$\phi(x) = f(x) + \sum_{s=1}^{\infty} \lambda^s \int_0^x K_s(x, y) f(y)\,dy \tag{3.16}$$

the series being convergent for all λ.

Again, it is sometimes possible to reduce the problem of the solution of a Volterra equation in more than one dimension to the solution of a differential equation with some associated boundary conditions. However, whether this can actually be done is a matter of luck.

Example 3.7

Solve the integral equation

$$\phi(x, y) = f(x, y) + \int_0^x \int_0^y \exp(x-\xi+y-\eta)\phi(\xi, \eta)\, \mathrm{d}\xi\, \mathrm{d}\eta$$

Now

$$K_1(x, y; \xi, \eta) = \exp(x-\xi+y-\eta)$$

$$K_2(x, y; \xi, \eta) = \int_\xi^x \int_\eta^y K(x, y; x', y')K(x', y'; \xi, \eta)\, \mathrm{d}x'\, \mathrm{d}y'$$

$$= \exp(x-\xi+y-\eta) \int_\xi^x \mathrm{d}x' \int_\eta^y \mathrm{d}y'$$

$$= \exp(x-\xi+y-\eta) \quad (x-\xi)(y-\eta)$$

Similarly

$$K_n(x, y; \xi, \eta) = \exp(x-\xi+y-\eta)\frac{(x-\xi)^{n-1}(y-\eta)^{n-1}}{[(n-1)!]^2}$$

and so

$$R(x, y); \xi, \eta) = -\sum_{n=1}^{\infty} K_n(x, y; \xi, \eta)$$

$$= -\sum_{n=0}^{\infty} \frac{(x-\xi)^n(y-\eta)^n}{(n!)^2} \exp(x-\xi+y-\eta)$$

and the solution is given by

$$\phi(x, y) = f(x, y) - \int_0^x \int_0^y R(x, y; \xi, \eta) f(\xi, \eta)\, \mathrm{d}\xi\, \mathrm{d}\eta$$

Alternatively, the problem may be put in differential equation form. It is clear that $\phi(0, 0) = f(0, 0)$. Differentiating the original equation with respect to x, it can be seen that

$$\begin{aligned}\frac{\partial\phi}{\partial x} &= \frac{\partial f}{\partial x} + \int_0^y \exp(y-\eta)\phi(x,\eta)\,\mathrm{d}\eta \\ &\quad + \int_0^x \int_0^y \exp(x-\xi+y-\eta)\phi(\xi,\eta)\,\mathrm{d}\xi\,\mathrm{d}\eta \\ &= \frac{\partial f}{\partial x} + \int_0^y \exp(y-\eta)\phi(x,\eta)\,\mathrm{d}\eta + \phi(x,y) - f(x,y) \qquad \text{(a)}\end{aligned}$$

It follows from this that

$$\frac{\partial\phi(x,0)}{\partial x} = \frac{\partial f(x,0)}{\partial x} + \phi(x,0) - f(x,0) \qquad \text{(b)}$$

is one of the boundary conditions. Similarly another one will be

$$\frac{\partial\phi(0,y)}{\partial y} = \frac{\partial f(0,y)}{\partial y} + \phi(0,y) - f(0,y) \qquad \text{(c)}$$

The integro-differential equation (a) can be differentiated again with respect to y and it follows that

$$\frac{\partial^2\phi}{\partial x\,\partial y} = \frac{\partial^2 f}{\partial x\,\partial y} + \phi + \int_0^y \exp(y-\eta)\phi(x,\eta)\,\mathrm{d}\eta + \frac{\partial\phi}{\partial y} - \frac{\partial f}{\partial y}$$

whence it follows that ϕ obeys the differential equation

$$\frac{\partial^2\phi}{\partial x\,\partial y} - \frac{\partial\phi}{\partial x} - \frac{\partial\phi}{\partial y} = \frac{\partial^2 f}{\partial x\,\partial y} - \frac{\partial f}{\partial x} - \frac{\partial f}{\partial y} + f \qquad \text{(d)}$$

and the solution of the problem is equivalent to that defined by the differential equation (d) together with the boundary conditions.

3.3 Convolution Type Kernels

If the kernel of a Volterra integral equation is of the form $K(x-y)$, the equation is said to be of convolution type and may be solved by use of the Laplace transform. The method of solution depends upon the well known result in Laplace transform theory (for details of this and other results in Laplace transform theory see Reference 4) that

$$\int_0^\infty \mathrm{e}^{-px} \int_0^x a(x-y)b(y)\,\mathrm{d}y\,\mathrm{d}x = \int_0^\infty \mathrm{e}^{-px} a(x)\,\mathrm{d}x \int_0^\infty \mathrm{e}^{-px} b(x)\,\mathrm{d}x$$

The quantity

$$\int_0^x a(x-y)b(y)\,\mathrm{d}y = \int_0^x a(y)b(x-y)\,\mathrm{d}y$$

is termed the convolution or faltung of the two functions $a(x)$ and $b(x)$. It will be convenient to note the Laplace transform of a function $a(x)$ by $\bar{a}$ and so on.

Consider now the integral of the first kind

$$f(x) = \int_0^x K(x-y)\phi(y)\,dy \tag{3.17}$$

It follows, on taking the Laplace transform, that

$$\bar{f} = \bar{K}\bar{\phi} \tag{3.18}$$

whence

$$\bar{\phi} = \bar{f}/\bar{K} \tag{3.19}$$

provided that the various transforms exist. The solution is given by finding the inverse transform of $\bar{\phi}$, that is the function $\phi(x)$ of which $\bar{\phi}$ is the transform. A theorem exists for this purpose, but sometimes the inverse transform can be seen immediately.

Example 3.8

Solve the integral equation

$$\int_0^x \sin\alpha(x-y)\phi(y)\,dy = 1-\cos\beta x$$

It will be observed that the equation is self-consistent. Taking the Laplace transform,

$$\frac{\alpha}{p^2+\alpha^2}\bar{\phi} = \frac{1}{p} - \frac{p}{p^2+\beta^2}$$

whence

$$\begin{aligned}\bar{\phi} &= \frac{\beta^2(p^2+\alpha^2)}{\alpha p(p^2+\beta^2)}\\ &= \frac{\alpha}{p} + \left(\frac{\beta^2-\alpha^2}{\alpha}\right)\frac{p}{p^2+\beta^2}\end{aligned}$$

and

$$\phi(x) = \alpha + \alpha^{-1}(\beta^2-\alpha^2)\cos\beta x$$

This integral equation could be solved equally well by reduction to a differential equation by differentiation.

Example 3.9

Solve the integral equation

$$\int_0^x \sin \alpha(x-y)\phi(y)\,dy = x$$

It will be observed that differentiating this equation with respect to x that

$$\alpha \int_0^x \cos \alpha(x-y)\phi(y)\,dy = 1$$

It can be seen that, on putting $x = 0$, there is an inconsistency. Nevertheless, it is still possible to follow through formally the Laplace transform processes which are involved in going from Eq. (3.19). Doing this, it follows that

$$\frac{\alpha}{p^2+\alpha^2}\bar{\phi} = \frac{1}{p^2}, \qquad \bar{\phi} = \alpha^{-1}(1+\alpha^2/p^2)$$

The inverse of $\bar{\phi}$ in this case is

$$\alpha^{-1}\delta(x)+\alpha x$$

Example 3.10

Solve the integral equation

$$\int_0^x \phi(x-y)[\phi(y)-2\sin ay]\,dy = x\cos ax$$

Taking the transform it follows that

$$\bar{\phi}\{\bar{\phi}-2a/(p^2+a^2)\} = (p^2-a^2)/(p^2+a^2)^2$$

whence

$$\bar{\phi} = (a \pm p)/(p^2+a^2)$$

and

$$\phi(x) = \sin ax \pm \cos ax$$

there being two possible solutions. It will be noted that although the equation is non-linear, it can in fact be solved by the methods of this section.

Example 3.11

Solve Abel's equation

$$\int_0^x (x-y)^{-\alpha}\phi(y) = f(x) \qquad 0 < \alpha < 1$$

The reason for the condition $\alpha < 1$ is so that the integral converges at its upper limit. The reason for the condition $\alpha > 0$ will emerge later. Using the result that

$$\int_0^\infty t^{z-1} \mathrm{e}^{-t} \mathrm{d}t = \Gamma(z)$$

the original equation transforms into

$$\Gamma(1-\alpha)p^{\alpha-1}\bar{\phi} = \bar{f}$$

whence

$$\bar{\phi} = \Gamma(1-\alpha)p^{1-\alpha}\bar{f}$$
$$= \frac{\sin \alpha\pi}{\pi} p(\Gamma(\alpha)p^{-\alpha})\bar{f}$$

whence

$$\phi(x) = \frac{\sin \alpha\pi}{\pi} \frac{\mathrm{d}}{\mathrm{d}x} \int_0^x (x-y)^{\alpha-1} f(y) \, \mathrm{d}y$$
$$= \frac{\sin \alpha\pi}{\pi} \int_0^x (x-y)^{\alpha-1} f'(y) \, \mathrm{d}y$$

It will be seen that the condition $\alpha > 0$ is necessary for these integrals to converge.

It is possible to solve inhomogeneous Volterra equations of the second kind with convolution kernels in exactly the same way. The equation

$$\phi(x) = f(x) + \int_0^x K(x-y)\phi(y) \, \mathrm{d}y$$

transforms into

$$\bar{\phi} = \bar{f} + \bar{K}\bar{\phi}$$

whence

$$\bar{\phi} = (1-\bar{K})^{-1}\bar{f} \tag{3.20}$$

and $\phi(x)$ may be found.

Example 3.12

Solve the integral equation

$$\phi(x) = x^3 + \int_0^x \mathrm{e}^{3(x-y)} \phi(y) \, \mathrm{d}y$$

It follows immediately that

$$\bar{\phi} = \frac{\Gamma(4)}{p^4} + \frac{1}{p-3}\bar{\phi}$$

whence

$$\begin{aligned}\bar{\phi} &= [1-(p-3)^{-1}]^{-1}\Gamma(4)p^{-4} \\ &= \left(1+\frac{1}{p-4}\right)\Gamma(4)p^{-4} \\ \phi(x) &= x^4 + \int_0^x e^{4(x-y)} y^4 \, dy\end{aligned}$$

Some integro-differential equations of suitable form can also be solved by this method. An example will make this clear.

Example 3.13

Solve the integro-differential equation

$$\phi''(x) + \int_0^x e^{2(x-y)} \phi'(y) = 1$$

$$\text{where} \quad \phi(0) = 0 \qquad \text{and} \qquad \phi'(0) = 0$$

Taking the Laplace transform, it follows that

$$p^2\bar{\phi} + p\bar{\phi}/(p-2) = p^{-1}$$

and

$$\bar{\phi} = p^{-2}(p-1)^{-2} = \frac{1}{(p-1)^2} - \frac{2}{p-1} + \frac{1}{p^2} + \frac{2}{p}$$

$$\phi(x) = x\,e^x - 2e^x + x + 2$$

It is also possible to solve simultaneous equations involving convolutions. Consider a system of n functions defined by the n equations

$$\phi_i(x) = f_i(x) + \sum_{j=1}^{n} \int_0^x K_{ij}(x-y)\phi_j(y) \, dy \tag{3.21}$$

The Laplace transform of this set of equations is

$$\bar{\phi}_i = \bar{f}_i + \sum_{j=1}^{n} \bar{K}_{ij}\bar{\phi}_j$$

There are thus n equations for n unknown functions $\bar{\phi}_j$.

Example 3.14

Solve the system of equations

$$\phi_1(x) = 1 + \int_0^x \phi_2(y)\,dy$$

$$\phi_2(x) = e^{2x} - \int_0^x e^{2(x-y)}\phi_1(y)\,dy$$

The transforms of these equations are

$$\bar{\phi}_1 = p^{-1}(1+\bar{\phi}_2)$$

$$\bar{\phi}_2 = (p-2)^{-1}(1-\bar{\phi}_1)$$

whence

$$\bar{\phi}_1 = (p-1)^{-1}, \qquad \bar{\phi}_2 = (p-1)^{-1}$$

and

$$\phi_1(x) = \phi_2(x) = e^x$$

Another type of equation which can be dealt with by the convolution result in Laplace transform theory is a homogeneous Volterra equation of the second kind of the form

$$x\phi(x) = \int_0^x K(x-y)\phi(y)\,dy \tag{3.22}$$

In this case the kernel is $x^{-1}K(x-y)$ and is singular, and the proof that there are neither eigenvalues nor eigenfunctions does not hold. Now

$$\int_0^\infty x\,e^{-px}\phi(x)\,dx = -\frac{d}{dp}\int_0^\infty e^{-px}\phi(x)\,dx$$

and so it follows that Eq. (3.22) is transformed into

$$-\frac{d\bar{\phi}}{dp} = \bar{K}\bar{\phi}$$

and it is necessary to integrate this equation. Clearly the arbitrary constant of integration arising here merely reflects the scale of ϕ. A solution will exist if

$$\int_0^x k(x-y)\phi(y)\,dy \quad \text{exists}$$

Example 3.15

Solve the integral equation

$$x\phi(x) = \lambda \int_0^x (1+a\mathrm{e}^{-\alpha y})\phi(x-y)\,\mathrm{d}y \qquad (3.23)$$

indicating for what ranges of a, α and λ a solution is possible. The transform of the original equation is

$$-\frac{\mathrm{d}\bar{\phi}}{\mathrm{d}p} = \lambda[p^{-1}+a(p+\alpha)^{-1}]\bar{\phi}$$

whence

$$\bar{\phi} = C\Gamma(\lambda)\Gamma(\lambda a)p^{-\lambda}(p+\alpha)^{-\lambda a}$$

where C is some arbitrary constant. The reason for the introduction of the gamma functions will be seen later. Let

$$u(x) = x^{\lambda-1}, \qquad \bar{u} = \Gamma(\lambda)p^{-\lambda}$$

$$v(x) = \mathrm{e}^{-\alpha x}x^{\lambda a-1}, \qquad \bar{v} = \Gamma(\lambda a)(p+\alpha)^{-\lambda a}$$

Thus $\bar{\phi} = \bar{u}\bar{v}$ and, apart from an arbitrary constant, the solution of the integral equation is given by

$$\phi(x) = \int_0^x u(y)v(x-y)\,\mathrm{d}y$$

It will be seen that for $u(x)$, $v(x)$ and $\phi(x)$ to be defined $\mathrm{Re}\,\lambda > 0$, $\mathrm{Re}\,\lambda a > 0$. Thus any value of λ with positive real part is a possible eigenvalue and the associated eigenfunction is given by $\phi(x)$.

Another type of integral equation which can be solved with the aid of the Laplace transform is the following

$$\phi(x) = f(x)+\lambda\int_x^\infty K(y-x)\phi(y)\,\mathrm{d}y \qquad x>0$$

The solution depends upon the result

$$\int_0^\infty \mathrm{e}^{-px}\,\mathrm{d}x\int_x^\infty K(y-x)\phi(y)\,\mathrm{d}y = \int_0^\infty K(u)\,\mathrm{e}^{pu}\,\mathrm{d}u\int_0^\infty \mathrm{e}^{-py}\phi(y)\,\mathrm{d}y$$

which, on putting $x = y+u$, gives

$$= \bar{K}(-p)\bar{\phi}(p)$$

This process is legitimate, provided that the domains of definition of $\bar{K}(-p)$ and $\bar{\phi}(p)$ in the p plane overlap. If this is so, then

$$\bar{\phi}(p) = \bar{f}(p)+\lambda\bar{K}(-p)\bar{\phi}(p)$$

and

$$\bar{\phi}(p) = \bar{f}(p)/[1-\lambda\bar{K}(-p)]$$

The function $\phi(x)$ which corresponds to $\bar{\phi}$ will be termed the principal solution because the solution may not be unique. This occurs when there is a non-trivial solution to the integral equation

$$\phi(x) = \lambda \int_x^\infty K(y-x)\phi(y)\,\mathrm{d}y \tag{3.24}$$

The reason that this may happen is as follows: let

$$\xi = x^{-1}, \quad \eta = y^{-1}, \quad K(y-x) = k(\xi,\eta), \quad \phi(x) = \Phi(\xi)$$

Then the integral equation (3.24) becomes

$$\Phi(\xi) = \lambda \int_0^\xi k(\xi,\eta)\eta^{-2}\Phi(\eta)\,\mathrm{d}\eta$$

The kernel of this is singular and so the condition for the non-appearance of eigensolutions is not satisfied.

Example 3.16

Solve the integral equation

$$\phi(x) = x+\lambda \int_x^\infty \mathrm{e}^{\alpha(x-y)}\phi(y)\,\mathrm{d}y \qquad x>0$$

It will be seen immediately that the solution need not be unique, for a solution of the integral equation

$$\phi(x) = \lambda \int_x^\infty \mathrm{e}^{\alpha(x-y)}\phi(y)\,\mathrm{d}y$$

is $\mathrm{e}^{(\alpha-\lambda)x}$, provided $\mathrm{Re}\,\lambda>0$. This may be termed the complementary function. The principal solution can be obtained as follows

$$K(x) = \mathrm{e}^{-\alpha x}, \qquad \bar{K}(p) = \frac{1}{p+\alpha}$$

and this is defined if $\mathrm{Re}\,(p+\alpha)>0$.

$\bar{K}(-p) = (\alpha-p)^{-1}$ which is defined if

$$\mathrm{Re}\,(\alpha-p)>0 \quad \text{i.e.} \quad \mathrm{Re}\,p<\mathrm{Re}\,\alpha \tag{a}$$

It follows that the Laplace transformation of the principal solution is given by

$$\bar{\phi} = p^{-2}+\lambda(\alpha-p)^{-1}\bar{\phi}$$

and

$$\bar{\Phi} = p^{-2}(p-\alpha+\lambda)^{-1}(p-\alpha)$$
$$= (\alpha-\lambda)^{-2}[\lambda p^{-1}+\alpha(\alpha-\lambda)p^{-2}-\lambda(p-\alpha+\lambda)^{-1}]$$

The principal solution is thus given by

$$(\alpha-\lambda)^{-2}[\lambda+\alpha(\alpha-\lambda)x-\lambda\, e^{(\alpha-\lambda)x}]$$

$\bar{\Phi}$ exists if the two conditions

$$\operatorname{Re} p > 0 \tag{b}$$

and

$$\operatorname{Re}(p-\alpha+\lambda) > 0 \quad \text{i.e.} \quad \operatorname{Re} p > \operatorname{Re}(\alpha-\lambda) \tag{c}$$

Conditions (a) and (b) can hold simultaneously if $\operatorname{Re}\alpha > 0$, when there exists a common domain $0 < \operatorname{Re} p < \operatorname{Re}\alpha$. Conditions (a) and (c) can hold simultaneously if

$$\operatorname{Re}\alpha > \operatorname{Re} p > \operatorname{Re}(\alpha-\lambda)$$

This can hold if

$$\operatorname{Re}\lambda > 0$$

Thus, if $\operatorname{Re}\lambda > 0$, there is a common domain of existence for $\bar{K}(-p)$ and $\bar{\Phi}(p)$ defined by

$$\operatorname{Re}\alpha > \operatorname{Re} p > \max\{\operatorname{Re}(\alpha-\lambda), 0\}$$

The conditions $\operatorname{Re}\alpha > 0$, $\operatorname{Re}\lambda > 0$ are in fact the conditions necessary for

$$\int_x^\infty e^{-\alpha y} y\, dy, \quad \int_x^\infty e^{-\alpha y} y\, dy$$

and

$$\int_x^\infty e^{-\alpha y} e^{(\alpha-\lambda)y}\, dy$$

to converge. It follows therefore that the solution of the integral equation is given by

$$(\alpha-\lambda)^{-2}[\lambda+\alpha(\alpha-\lambda)x]+C\, e^{(\alpha-\lambda)x}$$

where C is arbitrary. It will be noted that this solution is not valid if $\alpha = \lambda$. In this case

$$\phi(x) = \lambda \int_x^\infty e^{\lambda(x-y)}\phi(y)\, dy$$

is satisfied by $\phi(x) = 1$, provided that $\operatorname{Re}\lambda > 0$. The integral equation to be solved becomes

$$\phi(x) = x + \lambda \int_x^\infty e^{\lambda(x-y)} \phi(y) \, dy$$

In this case

$$\bar{K}(-p) = (\lambda - p)^{-1}$$

which is defined if

$$\operatorname{Re} p < \operatorname{Re} \lambda$$

The Laplace transformation of the principal solution is given by

$$\bar{\phi} = p^{-2} + \lambda(\lambda - p)^{-1} \bar{\phi}$$

and

$$\bar{\phi} = \frac{p - \lambda}{p^3}$$

giving

$$\phi(x) = x - \frac{\lambda x^2}{2}$$

$\bar{\phi}$ is defined for $\operatorname{Re} p > 0$.

Thus the solution is possible because p can be taken to lie within the domain $\operatorname{Re}\lambda > \operatorname{Re} p > 0$, and this is possible if $\operatorname{Re}\lambda > 0$.

The general solution is thus

$$\phi(x) = x - \frac{\lambda x^2}{2} + C$$

where C is arbitrary.

Example 3.17

Find the eigenfunctions of the integral equation

$$\phi(x) = \lambda \int_x^\infty \cos \alpha(x - y) \phi(y) \, dy$$

Whilst it is possible to solve this equation by reducing it to a differential equation, a more intuitive approach is also possible. Try a solution $e^{\gamma x}$ where γ is undetermined. Then

$$\int_x^\infty \cos \alpha(x-y) \phi(y) \, dy = \int_x^\infty \cos \alpha(x-y) e^{\gamma y} \, dy = -\frac{\gamma}{\gamma^2 + \alpha^2} e^{\gamma x}$$

Thus the integral equation is satisfied provided that

$$\lambda = -\gamma/(\gamma^2+\alpha^2)$$

which gives two possible values for γ if λ is given, or a value for λ when γ is given. The integral must converge however.

Now

$$\cos \alpha(x-y)\, e^{\gamma y} = \tfrac{1}{2}[e^{(\gamma - i\alpha)y}\, e^{i\alpha x} + e^{(\gamma + i\alpha)y}\, e^{-i\alpha x}]$$

and the condition for convergence is clearly given by

$$\operatorname{Re}(\gamma - i\alpha) < 0, \qquad \operatorname{Re}(\gamma + i\alpha) < 0$$

If α is real it is simply $\operatorname{Re}\gamma < 0$.

3.4. Some Miscellaneous Types of Volterra Integral Equations

(a) Equations Related to Abel's Equation

In this section will be considered some particular equations which can be solved by special techniques. Abel's equation has already been solved by the use of the Laplace transform, but can in fact be solved directly in the following manner:

If

$$f(x) = \int_0^x (x-y)^{-\alpha}\phi(y)\,dy \qquad 0 < \alpha < 1$$

$$\int_0^u \frac{f(x)\,dx}{(u-x)^{1-\alpha}} = \int_0^u \frac{dx}{(u-\alpha)^{1-\alpha}} \int_0^x \frac{\phi(y)\,dy}{(x-y)^{\alpha}}$$

$$= \int_0^u \phi(y)\,dy \int_y^u \frac{dx}{(u-x)^{1-\alpha}(x-y)^{\alpha}} \tag{3.25}$$

It can in fact be shown that the inner integral on the left hand side of Eq. (3.25) is equal to $\pi \operatorname{cosec} \pi\alpha$. Thus

$$\int_0^u \phi(y)\,dy = \frac{\sin \alpha\pi}{\pi} \int_0^u \frac{f(x)\,dx}{(u-x)^{1-\alpha}}$$

and

$$\phi(u) = \frac{\sin \alpha\pi}{\pi} \frac{d}{du} \int_0^u \frac{f(x)\,dx}{(u-x)^{1-\alpha}} \tag{3.26}$$

A more complicated equation which can be solved in the same way is

$$f(x) = \int_c^x \frac{\phi(y)\,dy}{\{p(x)-p(y)\}^\alpha} \qquad \begin{matrix} 0<\alpha<1 \\ c<x \end{matrix} \tag{3.27}$$

where $p(x)$ is a strictly monotonic increasing differentiable function with non-zero $p'(x)$ over some interval $x_0 \leqslant c < x < a$. Consider

$$\int_c^s \frac{p'(x)f(x)\,dx}{\{p(s)-p(x)\}^{1-\alpha}} = \int_c^s \int_c^x \frac{p'(x)\phi(y)\,dx\,dy}{\{p(s)-p(x)\}^{1-\alpha}\{p(x)-p(y)\}^\alpha}$$

$$= \int_c^s \phi(y)\,dy \left[\int_y^s \frac{p'(x)\,dx}{\{p(s)-p(x)\}^{1-\alpha}\{p(x)-p(y)\}^\alpha}\right]$$

$$= \pi \operatorname{cosec} \pi\alpha \int_c^s \phi(y)\,dy$$

It follows that

$$\phi(s) = \frac{\sin \alpha\pi}{\pi}\frac{d}{ds}\int_c^s \frac{p'(x)f(x)\,dx}{\{p(s)-p(x)\}^{1-\alpha}} \tag{3.28}$$

and this is the solution of the integral equation (3.28). It may be noted that this is applicable when the lower limit of integration c is replaced by $-\infty$. Also $f(c) = 0$ for consistency.

Example 3.18

Solve the integral equation

$$x = \int_0^x \frac{\phi(y)\,dy}{(x^2-y^2)^{\frac{1}{2}}}$$

The consistency condition is satisfied.

$$p(x) = x^2, \quad p'(x) = 2x \neq 0 \quad \text{if} \quad x > 0$$

Thus

$$\phi(y) = \pi^{-1} \sin\frac{\pi}{2}\frac{d}{dy}\int_0^y \frac{2x^2\,dx}{(y^2-x^2)^{\frac{1}{2}}}$$

$$= (2/\pi)\frac{d}{dy}\int_0^{\pi/2} y^2 \cos^2\theta\,d\theta$$

on making the transformation

$$x = y\cos\theta$$

$$\phi(y) = y$$

Example 3.19

Solve the integral equation

$$g(x) = x \int_0^x \frac{\phi(y)\,\mathrm{d}y}{(x^2 - y^2)^{\frac{1}{2}}} \qquad 0 \leqslant x$$

where

$$\begin{aligned} g(x) &= g_1(x) \qquad 0 \leqslant x < a \\ &= g_2(x) \qquad a \leqslant x \end{aligned}$$

and

$$\lim_{x \to a} g_1(x) \neq \lim_{x \to a} g_2(x)$$

Because $g(x)$ is defined through an integral, it will be continuous if ϕ is bounded. Because of the discontinuity, ϕ will contain a generalized function as well as an ordinary function. The formal solution is given by

$$\begin{aligned} \phi(y) &= \pi^{-1} \frac{\mathrm{d}}{\mathrm{d}y} \int_0^y \frac{2g(x)\,\mathrm{d}x}{(y^2 - x^2)^{\frac{1}{2}}} \\ &= (2/\pi) \frac{\mathrm{d}}{\mathrm{d}y} \int_0^{\pi/2} g(y \sin\theta)\,\mathrm{d}\theta \\ &= (2/\pi) \int_0^{\pi/2} g'(y \sin\theta) \sin\theta\,\mathrm{d}\theta \end{aligned}$$

Now it is possible to write

$$g(x) = g_1(x)\{1 - H(x-a)\} + g_2(x) H(x-a)$$

where

$$H(x) = 1 \qquad x \geqslant 0, \qquad = 0 \qquad x < 0$$

Thus

$$g'(x) = g_1'(x)[1 - H(x-a)] + g_2'(x) H(x-a) + \delta(x-a)\{g_2(x) - g_1(x)\}$$

Let

$$y \sin\alpha = a, \qquad \alpha = \operatorname{cosec}^{-1}(y/a)$$

Then

$$\begin{aligned}\phi(y) = (2/\pi)\int_0^{\pi/2} & g_1'(y\sin\theta)[1 - H(\overline{y\sin\theta - \sin\alpha})] \\ & + g_2'(y\sin\theta)H(\overline{y\sin\theta - \sin\alpha}) \\ & + \delta(\overline{y\sin\theta - \sin\alpha})[g_2(y\sin\theta) - g_1(y\sin\theta)]\sin\theta\,\mathrm{d}\theta \\ = (2/\pi) & \int_0^{\alpha} g_1'(y\sin\theta)\sin\theta\,\mathrm{d}\theta \\ & + (2/\pi)\int_\alpha^{\pi/2} g_2'(y\sin\theta)\sin\theta\,\mathrm{d}\theta \\ & + (2/\pi)[g_2(y\sin\alpha) - g_1(y\sin\alpha)](\tan\alpha)/y\end{aligned}$$

where the result

$$\int \delta\{f(\theta) - f(\alpha)\}\,\mathrm{d}\theta = \{f'(\alpha)\}^{-1}$$

has been used.

$$\begin{aligned}= (2/\pi)\int_0^{\alpha} & g_1'(y\sin\theta)\sin\theta\,\mathrm{d}\theta \\ & + \int_\alpha^{\pi/2} g_2'(y\sin\theta)\sin\theta\,\mathrm{d}\theta \\ & + [g_2(a) - g_1(a)]a/\{y\sqrt{(y^2 - a^2)}\}\end{aligned}$$

More discontinuities in g could be treated similarly.

Example 3.20

Solve the equation

$$\iint_D \frac{\phi(x, y)\,\mathrm{d}x\,\mathrm{d}y}{\{(y_0 - y)^2 - (x_0 - x)^2\}^{\frac{1}{2}}} = f(x_0, y_0)$$

when the domain D is the triangular area bounded by the straight lines $y = 0, \quad x - x_0 \pm (y - y_0) = 0$.

Although this integral equation involves two variables, it can be treated by the methods of this section, provided that a change of variable is made. The hint is given by the fact that

$$(y_0 - y)^2 - (x_0 - x)^2 = (y_0 + x_0 - y - x)(y_0 - x_0 - \overline{y - x})$$

Let

$$\xi = x+y, \quad \eta = y-x, \quad \xi_0 = x_0+y_0, \quad \eta_0 = y_0-x_0.$$

Then

$$\frac{\partial(\xi,\eta)}{\partial(x,y)} = \tfrac{1}{2}, \quad \frac{\partial(x,y)}{\partial(\xi,\eta)} = 2$$

$$y = 0 \text{ transforms into } \xi+\eta = 0$$

$$x-x_0+y-y_0 = 0 \text{ into } \eta = \eta_0$$

and

$$x-x_0-y+y_0 = 0 \text{ into } \xi = \xi_0$$

Thus

$$F(\xi_0,\eta_0) = \int_D\int \frac{\phi(x,y)\,\mathrm{d}x\,\mathrm{d}y}{\{(y_0-y)^2-(x_0-x)^2\}^{\frac{1}{2}}} = 2\int_\Delta\int \frac{\Phi(\xi,\eta)\,\mathrm{d}\xi\,\mathrm{d}\eta}{[(\xi_0-\xi)(\eta_0-\eta)]^{\frac{1}{2}}}$$

where Δ is the triangular area bounded by the three straight lines

$$\xi+\eta = 0, \quad \xi = \xi_0, \quad \eta = \eta_0$$

and

$$\Phi(\xi,\eta) = \phi(x,y)$$

and

$$F(\xi_0,\eta_0) = f(x_0,y_0)$$

Let Δ_0 be the triangular domain in the ξ_0,η_0 plane defined by the three straight lines

$$\xi_0+\eta_0 = 0, \quad \xi_0 = \alpha, \quad \eta_0 = \beta$$

Thus

$$\tfrac{1}{4}\int_{\Delta_0}\int \frac{\mathrm{d}\xi_0\,\mathrm{d}\eta_0}{[(\alpha-\xi_0)(\beta-\eta_0)]^{\frac{1}{2}}}\int_\Delta\int \frac{\Phi(\xi,\eta)\,\mathrm{d}\xi\,\mathrm{d}\eta}{[(\xi_0-\xi)(\eta_0-\eta)]^{\frac{1}{2}}} = 2\int_{\Delta_0}\int \frac{F(\xi_0,\eta_0)\,\mathrm{d}\xi_0\,\mathrm{d}\eta_0}{[(\alpha-\xi_0)(\beta-\eta_0)]^{\frac{1}{2}}}$$

It can be verified that the quadruple integral may be rewritten

$$\tfrac{1}{4}\int_0^\alpha\int_0^\beta \mathrm{d}\xi\,\mathrm{d}\eta\Phi(\xi,\eta)\int_\xi^\alpha \frac{\mathrm{d}\xi_0}{[(\alpha-\xi_0)(\xi_0-\xi)]^{\frac{1}{2}}}\int_\eta^\beta \frac{\mathrm{d}\eta_0}{[(\beta-\eta_0)(\eta_0-\eta)]^{\frac{1}{2}}}$$

$$= \frac{\pi^2}{4}\int_0^\alpha \mathrm{d}\xi\int_0^\beta \mathrm{d}\eta\Phi(\xi,\eta)$$

It follows that

$$\Phi(\alpha, \beta) = (8/\pi^2)\frac{\partial^2}{\partial\alpha\,\partial\beta}\int_{\Delta_0}\int \frac{F(\xi_0, \eta_0)\,d\xi_0\,d\eta_0}{[(\alpha-\xi_0)(\beta-\eta_0)]^{\frac{1}{2}}}$$

Let $\alpha = u+v$, $\beta = v-u$.

Then

$$\frac{\partial^2}{\partial\alpha\,\partial\beta} = \tfrac{1}{4}\left(\frac{\partial^2}{\partial v^2}-\frac{\partial^2}{\partial u^2}\right)$$

and it can be seen that

$$\phi(u, v) = \pi^{-2}\left[\frac{\partial^2\psi}{\partial v^2}-\frac{\partial^2\psi}{\partial u^2}\right]$$

where

$$\psi(u, v) = \int_{D_0}\int \frac{f(x_0, y_0)\,dx_0\,dy_0}{[(v-y_0)^2-(u-x_0)^2]^{\frac{1}{2}}}$$

where the domain D_0 is defined in the x_0, y_0 plane by the three straight lines $y_0 = 0$, $y_0 - v \pm (x_0 - u) = 0$. The transformation from α, β to u, v is the exact reverse of the transformation from (x, y) to (ξ, η).

Example 3.21

Solve the integral equation

$$\int_0^x (x-y)^{\alpha-1}\phi(y)\,dy = x^{\beta} \qquad \alpha > 1, \quad \beta \geqslant 0$$

It is true that this equation can be solved by the Laplace transform, but an intuitive approach can lead to a solution by inspection. If α is integral,

$$\int_0^x (x-y)^{\alpha-1}\phi(y)\,dy$$

is, apart from a constant factor, the αth integral of ϕ (Appendix A) and the αth integral of x^p is, apart from a constant factor $x^{\alpha+p}$. Thus, if α is integral, the solution of the integral equation is, apart from a constant factor $x^{\beta-\alpha}$ as $\alpha+p = \beta$.

The hint is therefore given to look for a solution of the form $Kx^{\beta-\alpha}$. If a solution exists, it will be unique. Then

$$K\int_0^x (x-y)^{\alpha-1}y^{\beta-\alpha}\,dy = x^{\beta}$$

Putting

$$y = x\tau$$

$$Kx^{\beta} \int_0^1 (1-\tau)^{\alpha-1} \tau^{\beta-\alpha} \, d\tau = x^{\beta}$$

whence

$$K = [B(\alpha-1, \beta-\alpha+1)]^{-1}$$

where B is the beta function.

(b) Non-linear Volterra Equations

In Section 3.2, a discussion was given of the possibility of an iterative solution for Volterra type equations which were linear in the unknown function ϕ. These ideas can be carried over to the solution of non-linear Volterra equations of the second kind, by the process known in differential equation theory as Picard's method. Consider the solution of the differential equation

$$\frac{d\phi}{dx} = g(x, \phi) \tag{3.29}$$

subject to the condition that at $x = a$, $\phi = b$. The solution of the differential equation and its associated boundary conditions is equivalent to the integral equation

$$\phi(x) = b + \int_a^x g\{x', \phi(x')\} \, dx' \tag{3.30}$$

The solution is carried out by the forming of the sequence of functions $\phi_n(x)$ defined by

$$\phi_0(x) = b$$

$$\phi_n(x) = b + \int_a^x g\{x', \phi(x')\} \, dx' \tag{3.31}$$

Consider now the Volterra equation of the second kind

$$\phi(x) = f(x) + \lambda \int_0^x F\{x, y, \phi(y)\} \, dy \tag{3.32}$$

Equation (3.30) is a special case of Eq. (3.32). Construct a sequence of functions $\phi_n(x)$ defined as follows:

$$\phi_0(x) = f(x)$$

$$\phi_n(x) = f(x) + \lambda \int_a^x F\{x, y, \phi_{n-1}(y)\} \, dy \tag{3.33}$$

It can be seen that, because of the fact that F is non-linear in ϕ, this process can be applied to the case where $f(x)$ is zero, and that it is possible for a sequence to be generated which does not tend to zero. In this case, the function to which the sequence of function tends in the limit is an eigenfunction corresponding to the eigenvalue λ. There may of course be, because of the non-linearity, more than one eigenfunction, but because of the nature of the sequence process, only one will turn up.

Suppose however that the general case of non-zero $f(x)$ is considered, and λ is absorbed into F. Two questions are to be considered. Firstly, does the sequence converge and secondly, if it does converge, does it converge to the solution of the integral equation? It will be proved that the answers to these two questions are in the affirmative over some domain

$$|x-a| \leqslant \alpha, |y-a| \leqslant \alpha, |\phi - f(a)| \leqslant \beta$$

α and β being positive constants, provided that

$$|F| < M \tag{3.34}$$

and

$$|F(x, y, \phi_A) - F(x, y, \phi_B)| < K|\phi_A - \phi_B| \tag{3.35}$$

where M and K are positive constants and x, y, ϕ_A and ϕ_B lie within the ranges of values just specified. Now

$$\phi_1(x) = f(x) + \int_a^x F\{x, y, f(y)\}\,dy$$

and so

$$\begin{aligned} |\phi_1(x) - \phi_0(x)| &= \left| \int_a^x F\{x, y, f(y)\}\,dy \right| \\ &\leqslant |x-a|M \leqslant M\alpha \end{aligned} \tag{3.36}$$

also

$$\begin{aligned} |\phi_2(x) - \phi_1(x)| &= \left| \int_a^x [F\{x, y, \phi_1(y)\} - F\{x, y, \phi_0(y)\}]\,dy \right| \\ &\leqslant \int_a^x |F\{x, y, \phi_1(y)\} - F\{x, y, \phi_0(y)\}|\,dy \\ &\leqslant \int_a^x K|\phi_1(y) - \phi_0(y)|\,dy \\ &\leqslant \int_0^x KM|y-a|\,dy = KM\frac{(x-a)^2}{2} \\ &\leqslant KM\frac{\alpha^2}{2} \end{aligned}$$

Similarly

$$|\phi_n(x) - \phi_{n-1}(x)| \leqslant \frac{1}{n!} MK^{n-1}\alpha^n \tag{3.37}$$

Thus the sum of the infinite series

$$|\phi_0(x)| + \sum_{n=1}^{\infty} |\phi_n(x) - \phi_{n-1}(x)| \leqslant M + \sum_{n=1}^{\infty} \frac{1}{n!} MK^{n-1}\alpha^n = K^{-1}M\,\mathrm{e}^{K\alpha}$$

and so the series is absolutely convergent.

It follows therefore that the sequence

$$\phi_m(x) = \phi_0(x) + \sum_{n=1}^{m} \{\phi_n(x) - \phi_{n-1}(x)\} \tag{3.38}$$

is absolutely convergent because each of the terms $\phi_n(x) - \phi_{n-1}(x)$ is dominated by the terms of the sequence $|\phi_n(x) - \phi_{n-1}(x)|$ which converges absolutely. Thus the first question has been answered affirmatively.

There remains the question of proving that the sequence $\phi_n(x)$ actually does converge to the solution of the integral equation. Let

$$\phi^*(x) = \lim_{m \to \infty} \phi_m(x) = \phi_0(x) + \sum_{n=1}^{\infty} \{\phi_n(x) - \phi_{n-1}(x)\}$$

The functions $\phi_n(x)$ are continuous, and $\phi^*(x)$ is defined by an absolutely convergent series of continuous functions $\phi_n(x) - \phi_{n-1}(x)$. The series is also, as can be seen from the proof of convergence given previously, uniformly convergent, and so $\phi^*(x)$ is continuous and $\phi^*(x) - \phi_m(x)$ tends uniformly to zero as m increases. Furthermore,

$$|F\{x, y, \phi^*(y)\} - F\{x, y, \phi_m(y)\}| < K|\phi^* - \phi_m|$$

and so

$$\left| \int_a^x [F\{x, y, \phi^*(y)\} - F\{x, y, \phi_m(y)\}]\,\mathrm{d}y \right|$$

will tend uniformly to zero.

Thus the limit of the equation

$$\phi_m(x) = f(x) + \int_a^x F\{x, y, \phi_{m-1}(y)\}\,\mathrm{d}y \tag{3.39}$$

is

$$\phi^*(x) = f(x) + \int_a^x F\{x, y, \phi^*(y)\}\,\mathrm{d}y$$

and so $\phi^*(x)$ is the required solution of the integral equation.

For the case of $f(x)$ zero, the first approximation to the eigenfunction sequence is given by

$$\phi_1(x) = \lambda \int_a^x F(x, y, 0)\,\mathrm{d}y$$

The rest of the process continues as before.

Example 3.22

Solve the integral equation

$$\phi(x) = \int_0^x \frac{1+\phi(y)}{1+y}\,\mathrm{d}y$$

It will be noted that this equation is not meaningful if $x \leqslant -1$. A sequence of functions is generated as follows

$$\phi_0(x) = 0$$

$$\phi_1(x) = \int_0^x \frac{\mathrm{d}y}{1+y} = \log(1+x)$$

$$\phi_2(x) = \int_0^x \frac{1+\log(1+y)\,\mathrm{d}y}{(1+y)}$$

$$= \log(1+x) + \frac{[\log(1+x)]^2}{2}$$

The general term in the sequence is given by

$$\phi_m(x) = \sum_{s=1}^{m} \frac{[\log(1+x)]^s}{s!}$$

and

$$\phi^*(x) = \lim_{m\to\infty} \phi_m(x) = \sum_{s=1}^{\infty} \frac{[\log(1+x)]^s}{s!}$$

$$= \exp\{\log(1+x)\} - 1 = x$$

It can be verified that this is indeed the solution. It is worth noting that the convergence conditions are satisfied because in

$$|x| \leqslant \alpha, \quad |y| \leqslant \alpha, \quad |\phi| \leqslant \beta$$

$$\left|\frac{1+\phi}{1+y}\right| \leqslant \frac{1+\beta}{1-\alpha} \qquad \alpha \leqslant 1-\delta < 1$$

and

$$\left|\frac{1+\phi_A}{1+y}-\frac{1+\phi_B}{1+y}\right| \leqslant \frac{1}{1-\alpha}\left|\phi_A-\phi_B\right|$$

and so if $|x| \leqslant 1-\delta < 1$, the convergence conditions are satisfied.

It can easily be seen that the solution of

$$\phi(x) = \lambda \int_0^x \frac{1+\phi(y)}{1+y}\,\mathrm{d}y \qquad \operatorname{Re}\lambda > 0$$

is

$$\phi(x) = (1+x)^{\lambda}-1$$

EXERCISES

1. Solve the integral equation for $\phi(x)$

$$\phi(x) = \mathrm{e}^{-\alpha x}+\lambda \int_0^x \mathrm{e}^{-\beta(x-y)}\phi(y)\,\mathrm{d}y \qquad 0 \leqslant x$$

distinguishing between the cases $\lambda \neq \beta-\alpha$ and $\lambda = \beta-\alpha$. (Wales)

2. Solve the integral equation

$$\phi(x) = 2\cos ax+\int_0^x (x-t)\phi(t)\,\mathrm{d}t \qquad x \geqslant 0$$ (Wales)

3. Prove that the solution of the integral equation

$$\phi(x) = \lambda \int_0^x [A(x)/A(y)]\phi(y)\,\mathrm{d}y+f(x) \qquad x \geqslant 0$$

is given by

$$\phi(x) = f(x)+\lambda \int_0^x [A(x)/A(y)]\,\mathrm{e}^{\lambda(x-y)}f(y)\,\mathrm{d}y$$

provided that $A(x)$ is non-zero for positive x.

4. Solve the integral equation

$$\int_0^x \exp k(x-y)\phi(y)\,\mathrm{d}y = \sin\beta x \qquad x \geqslant 0$$

5. Solve the integral equation

$$\int_0^x \cosh 3(x-y)\phi(y)\,\mathrm{d}y = 1-\alpha\cos\alpha x \qquad x \geqslant 0$$

distinguishing between the cases $\alpha \neq 1$ and $\alpha = 1$.

6. Prove that the solutions of the integral equation

$$\int_0^x \phi(x-y)[\phi(y)-2]\,dy = x(e^{-ax}-1) \qquad x \geqslant 0$$

are $1 \pm e^{-ax}$.

7. Solve the integral equation

$$\int_0^x (x-y)^{\frac{1}{2}}\phi(y)\,dy = x \qquad x \geqslant 0$$

8. Solve the integral equation

$$\int_0^x \cos a(x-y)\phi(y)\,dy = \sin ax \qquad x \geqslant 0$$

9. Solve the integral equation discussed in Section 1.2b

$$A = AK(t) + \int_0^t K(t-\tau)\phi(\tau)\,d\tau \qquad t > 0$$

when the proportion of unsold goods $K(t)$ is given by

(a) $\exp(-\alpha t)$ (b) $1-\alpha t \quad t \leqslant \alpha^{-1}; \quad = 0 \quad t > \alpha^{-1}$

Explain why the consistency condition would be expected to be satisfied in this problem.

10. Let $p(x)$ be a positive monotonic increasing differentiable function with a non-zero derivative in some interval $a < x < b$. Show that the solution of the integral equation

$$f(x) = \int_x^b \frac{\phi(y)\,dy}{\{p(y)-p(x)\}^{\alpha}} \qquad a \leqslant x \leqslant b, \quad 0 < \alpha < 1$$

is

$$\phi(y) = -\pi^{-1}\sin\alpha\pi \int_y^b \frac{p'(z)f(z)\,dz}{\{p(z)-p(y)\}^{1-\alpha}}$$

11. Prove that, if

$$f(x) = x\int_x^{\infty} \frac{g(y)\,dy}{(y^2-x^2)^{\frac{1}{2}}} \qquad x \geqslant 0$$

then

$$g(y) = -(2/\pi)\int_0^{\infty} \cosh z\, f'(y\cosh z)\,dz$$

12. Show that the eigenfunctions of the integral equation

$$\phi(x) = \lambda \int_x^\infty \frac{(t-x)^{n-1}}{(n-1)!} \phi(t)\,dt$$

are $\exp\{-\lambda^{1/n}x\}$ and state what the range of eigenvalues is.

13. Show that a possible solution of

$$\phi(x) = \lambda \int_0^x (t^{-1}-x^{-1})\phi(t)\,dt \qquad x \geqslant 0$$

is given by $\phi(x) = x^\alpha$ and find the corresponding value of λ, and the possible ranges of α and λ.

14. Solve the integral equation

$$x \int_x^\infty \frac{\phi(y)\,dy}{(y^2-x^2)^{\frac{1}{2}}} \begin{matrix} = x^{-1} & \quad 0 \leqslant x < a \\ = 2x^{-1} & \quad a \leqslant x \end{matrix}$$

15. Using the iterative method solve the integral equation

$$\phi(x) = \int_0^x \frac{1+[\phi(y)]^2}{1+y^2}\,dy \qquad x \geqslant 0$$

16. Solve the following system of integral equations:

$$\phi_1(x) = 1 - \int_0^x \phi_1(y)\,dy + 4\int_0^x e^{x-y}\phi_2(y)\,dy$$

$$\phi_2(x) = e^x - \int_0^x e^{y-x}\phi_1(y)\,dy + \int_0^x \phi_2(y)\,dy \qquad x \geqslant 0$$

17. Solve the integral equation

$$\phi(x) = \sin x + \cos x + 2\int_x^\infty \cos(x-y)\phi(y)\,dy$$

and verify that the solution obtained does in fact satisfy the equation.

18. Find the eigenfunctions of the integral equation

$$\phi(x) = \lambda \int_x^\infty \sin\alpha(x-y)\phi(y)\,dy$$

and give the conditions on λ and α which must be satisfied for them to exist.

4 Integral Equations and Transformations

4.1 Preliminary

Suppose that T is some linear operator which, when it acts on a function F defined over some domain D, gives rise to a function f which is defined on some domain Δ.

Formally,

$$f = TF \tag{4.1}$$

Suppose that S is some other linear operator which, when acting on the function f, defined in the domain Δ transforms it into the function F defined over a domain D.

Formally,

$$F = Sf \tag{4.2}$$

The process represented in Eq. (4.1) is termed a transformation, and the process represented in Eq. (4.2) is also a transformation. Equations (4.1) and (4.2) may be combined to give

$$f = TSf \quad \text{and} \quad F = STF \tag{4.3}$$

For this reason it is possible to term T the inverse transformation of S and vice versa.

The application to integral equations is as follows: When T is some integral process (an integral transform) then the solution of the integral equation (4.1) for F is given by Eq. (4.2), that is by the inverse transformation. Such an inverse transformation need not give an unique result, as there may exist functions G in D, such that $TG = 0$.

In this chapter, a number of illustrations will be given of these processes using various transformations. It will be assumed always that the functions considered satisfy the appropriate conditions for the operators to be meaningful. Detailed proofs and their justifications will not in general be given and for statements about the formulation of the integral transforms and their inverses see Reference 4.

Example 4.1

Let F be some quantity which is defined over some field. Let T be a transformation which transforms all such F into an entity TF which is defined over the same field. Let T be such that

$$T^2F = T(TF) = F$$

for all F. Solve formally the equations:

(a) $$\phi = \lambda T\phi \qquad \lambda \text{ is a number}$$

(b) $$\phi = \lambda T\phi + F$$

where ϕ is also defined over the field.

Let

$$\phi = u + \mu Tu$$

where μ is an undetermined number and u is some quantity defined over the field.

In case (a)

$$u + \mu Tu = \lambda T(u + \mu Tu) = \lambda\mu u + \lambda Tu$$

A solution is given by

$$\mu = \lambda, \quad \lambda\mu = 1, \quad u \text{ arbitrary}$$

and so a formal solution is given by

$$\lambda = \pm 1, \quad \phi = u \pm Tu$$

provided that the operations Tu, $T(Tu) = T^2u$ are legitimate.

In case (b)

$$T\phi = \lambda T^2\phi + \lambda TF = \lambda\phi + \lambda TF$$

whence

$$\phi(1 - \lambda^2) = f + \lambda Tf$$

provided that $\lambda^2 \neq 1$, and the operations are legitimate.

4.2 Fourier Integral Equations

If $f(x)$ is a continuous function, then

$$f(x) = \frac{1}{\sqrt{(2\pi)}} \int_{-\infty}^{\infty} e^{i\omega x} F(\omega)\, d\omega \tag{4.4a}$$

where

$$F(\omega) = \frac{1}{\sqrt{(2\pi)}} \int_{-\infty}^{\infty} e^{-i\omega x'} f(x') \mathrm{d}x' \tag{4.4b}$$

Equation (4.4b) gives the solution of the integral Eq. (4.4a) for F and vice versa. If $f(x)$ is real, it may be seen, using the odd property of $\sin \omega x$ and the even property of $\cos \omega x$, that the following results hold:

If

$$f(x) = \left(\frac{2}{\pi}\right)^{\frac{1}{2}} \int_0^{\infty} \cos \omega x \phi(\omega) \mathrm{d}\omega \qquad 0 \leqslant x \tag{4.5a}$$

then

$$\phi(\omega) = \left(\frac{2}{\pi}\right)^{\frac{1}{2}} \int_0^{\infty} \cos \omega x f(x) \mathrm{d}x \qquad 0 \leqslant \omega \tag{4.5b}$$

$\phi(\omega)$ and $f(x)$ are the cosine transforms of one another.

If

$$f(x) = \left(\frac{2}{\pi}\right)^{\frac{1}{2}} \int_0^{\infty} \sin \omega x \phi(\omega) \mathrm{d}\omega \qquad 0 \leqslant x \tag{4.6a}$$

then

$$\phi(\omega) = \left(\frac{2}{\pi}\right)^{\frac{1}{2}} \int_0^{\infty} \sin \omega x f(x) \mathrm{d}x \qquad 0 \leqslant \omega \tag{4.6b}$$

$\phi(\omega)$ and $f(x)$ are the sine transforms of one another. It will be seen that both of these transforms satisfy the conditions of Example 4.1.

Example 4.2

Solve the integral equation

$$\frac{a}{a^2 + x^2} = \int_0^{\infty} \cos \omega x \phi(\omega) \mathrm{d}\omega \qquad (a > 0)$$

$$\phi(\omega) = \frac{2}{\pi} \int_0^{\infty} \frac{a \cos \omega x}{a^2 + x^2} \mathrm{d}x$$

$$= \frac{1}{2\pi i} \int_{-\infty}^{\infty} \frac{2ia\, e^{i\omega x} \mathrm{d}x}{a^2 + x^2}$$

because $\sin \omega x$ is odd in x.

Evaluation of the integral by the methods of the complex integral calculus gives

$$\phi(\omega) = e^{-\omega a}, \quad \omega > 0$$

Example 4.3

Solve the integral equation

$$\phi(x) = \lambda \int_0^\infty \cos \omega x \phi(\omega) \mathrm{d}\omega$$

$\phi(x)$ is an even function of x.

It will be sufficient to solve for positive x and replace this by x in order to obtain the solution for all x.

Because the inverse of a cosine transformation is another cosine transformation, the following process is suggested. Look for a solution of the form

$$\phi(x) = U(x) \pm V(x)$$

where

$$V(x) = \left(\frac{2}{\pi}\right)^{\frac{1}{2}} \int_0^\infty \cos \omega x U(\omega) \mathrm{d}\omega$$

Thus

$$U(x) \pm \left(\frac{2}{\pi}\right)^{\frac{1}{2}} \int_0^\infty \cos \omega x U(\omega) \mathrm{d}\omega$$

$$= \lambda \int_0^\infty \cos \omega x \left[U(\omega) \pm \left(\frac{2}{\pi}\right)^{\frac{1}{2}} \int_0^\infty \cos \omega t U(t) \mathrm{d}t \right] \mathrm{d}\omega$$

$$= \lambda \int_0^\infty \cos \omega x U(\omega) \mathrm{d}\omega \pm \left(\frac{\pi}{2}\right)^{\frac{1}{2}} \lambda U(x)$$

This is true if $\lambda = \pm(2/\pi)^{\frac{1}{2}}$.

Thus to $\lambda = (2/\pi)^{\frac{1}{2}}$, there corresponds a solution $U(x)+V(x)$ and to $\lambda = -(2/\pi)^{\frac{1}{2}}$ there corresponds a solution $U(x)-V(x)$. These solutions will be valid, provided that all the integrals exist; U is arbitrary. Thus in this case to the two eigenvalues $\lambda = \pm(2/\pi)^{\frac{1}{2}}$ there exist an infinity of eigenfunctions.

Example 4.4

Solve the integral equation

$$\phi(x) = f(x) + \lambda \left(\frac{2}{\pi}\right)^{\frac{1}{2}} \int_0^\infty \cos xy \phi(y) \mathrm{d}y$$

If $\lambda = \pm 1$ there will not in general be any solution. This follows from Example 4.3. Look for a solution similar to that obtained in Example 4.3. This may be done by taking the transform of the original equation

$$\left(\frac{2}{\pi}\right)^{\frac{1}{2}} \int_0^\infty \cos xy \phi(y) \mathrm{d}y = \left(\frac{2}{\pi}\right)^{\frac{1}{2}} \int_0^\infty \cos xy f(y) \mathrm{d}y + \lambda \phi(x)$$

It follows that

$$(1-\lambda^2)\phi(x) = f(x)+\lambda\left(\frac{2}{\pi}\right)^{\frac{1}{2}}\int_0^\infty \cos xyf(y)\mathrm{d}y$$

and this solution is valid provided that the integrals converge. Now if $1-\lambda^2 = 0$, and $f(x)$ is a function such that

$$f(x)+\lambda\left(\frac{2}{\pi}\right)^{\frac{1}{2}}\int_0^\infty \cos xyf(y)\mathrm{d}y = 0$$

it follows that $\phi(x)$ can be any function for which the integrals converge.

Example 4.5

Solve the integral equation

$$\phi(x) = \lambda\int_{-\infty}^{\infty} \mathrm{e}^{i\omega x}\phi(\omega)\mathrm{d}\omega$$

Let

$$\phi(x) = U(x)\pm V(x)$$

$$V(x) = \frac{1}{\sqrt{(2\pi)}}\int_{-\infty}^{\infty} \mathrm{e}^{i\omega x}U(\omega)\mathrm{d}\omega$$

Thus

$$U(x)\pm\frac{1}{\sqrt{(2\pi)}}\int_{-\infty}^{\infty} \mathrm{e}^{i\omega x}U(\omega)\mathrm{d}\omega = \lambda\int_{-\infty}^{\infty} \mathrm{e}^{i\omega x}U(\omega)\mathrm{d}\omega$$
$$\pm\frac{\lambda}{\sqrt{(2\pi)}}\int_{-\infty}^{\infty} \mathrm{e}^{i\omega x}\mathrm{d}\omega\int_{-\infty}^{\infty} \mathrm{e}^{i\omega y}U(y)\mathrm{d}y$$

If $\lambda = \pm\dfrac{1}{\sqrt{(2\pi)}}$, this can be written as

$$U(x) = \frac{1}{2\pi}\int_{-\infty}^{\infty} \mathrm{e}^{i\omega x}\mathrm{d}\omega\int_{-\infty}^{\infty} \mathrm{e}^{i\omega y}U(y)\mathrm{d}y \tag{A}$$

This however is not the Fourier integral formula. Nevertheless, if U is even,

$$\int_{-\infty}^{\infty} \mathrm{e}^{i\omega y}U(y)\mathrm{d}y = \int_{-\infty}^{\infty} (\mathrm{e}^{-i\omega y}+2i\sin\omega y)U(y)\mathrm{d}y$$
$$= \int_{-\infty}^{\infty} \mathrm{e}^{-i\omega y}U(y)\mathrm{d}y$$

and so Eq. (A) is satisfied by any even function U. Thus the eigenvalues are $\pm1/\sqrt{(2\pi)}$ and they are associated with eigenfunctions $U(x)\pm V(x)$.

Example 4.6

Solve the integral equation

$$\phi(x) = f(x) + \lambda \int_{-\infty}^{\infty} k(x-y)\phi(y)\mathrm{d}y$$

Let capital letters denote Fourier transforms; e.g.

$$F(\omega) = \frac{1}{\sqrt{(2\pi)}} \int_{-\infty}^{\infty} \mathrm{e}^{i\omega x} f(x)\mathrm{d}x$$

Then

$$\Phi(\omega) = F(\omega) + \sqrt{(2\pi)}\lambda K(\omega)\Phi(\omega)$$

and

$$\Phi(\omega) = \left[1 + \frac{\sqrt{(2\pi)}\lambda K(\omega)}{1 - \sqrt{(2\pi)}\lambda K(\omega)}\right] F(\omega)$$

Let

$$l(x) = \frac{1}{\sqrt{(2\pi)}} \int_{-\infty}^{\infty} \frac{\mathrm{e}^{-i\omega x} K(\omega)}{1 - \sqrt{(2\pi)}\lambda K(\omega)} \mathrm{d}\omega$$

and

$$\phi(x) = f(x) + \int_{-\infty}^{\infty} l(x-y) f(y)\mathrm{d}y$$

It must be realized that this solution is only the principal solution. Under certain circumstances a complementary function may exist. Consider

$$\begin{aligned}\phi(x) &= \lambda \int_{-\infty}^{\infty} k(x-y)\phi(y)\mathrm{d}y \\ &= \lambda \int_{-\infty}^{\infty} k(u)\phi(x-u)\mathrm{d}u\end{aligned}$$

A solution is clearly given by $\mathrm{e}^{\alpha x}$ provided that

$$1 = \lambda \int_{-\infty}^{\infty} k(u)\mathrm{e}^{-\alpha u}\mathrm{d}u$$

If α is given and the integral converges, this defines an eigenvalue λ. If the integral does not converge, there is not an eigenvalue and so there is no corresponding eigenfunction. Alternatively, if λ is given, there is a (probably) transcendental equation for α and there can be more than one eigenfunction. It may be noted that an equation of the form

$$\phi(x) = f(x) + \int_{0}^{\infty} k(x/y)\phi(y)\mathrm{d}y/y$$

can be transformed into an equation of the form discussed by the changes of variable $x = \mathrm{e}^{\xi}$, $y = \mathrm{e}^{\eta}$.

Example 4.7

Solve the integral equation

$$\phi(x) = \lambda \int_0^\infty \frac{\phi(y)\mathrm{d}y}{x+y}$$

Let $x = \mathrm{e}^\xi$, $y = \mathrm{e}^\eta$, $\psi(\xi) = \mathrm{e}^{\frac{1}{2}\xi}\phi(x)$

$$\begin{aligned}\psi(\xi) &= \frac{\lambda}{2}\int_{-\infty}^{\infty} \operatorname{sech}\tfrac{1}{2}(\xi-\eta)\psi(\eta)\mathrm{d}\eta \\ &= \frac{\lambda}{2}\int_{-\infty}^{\infty} \operatorname{sech}\tfrac{1}{2}\eta\psi(\xi-\eta)\mathrm{d}\eta\end{aligned}$$

Let

$$\psi(\xi) = \mathrm{e}^{\alpha\xi}$$

$$1 = \frac{\lambda}{2}\int_{-\infty}^{\infty} \operatorname{sech}\tfrac{1}{2}\eta\mathrm{e}^{-\alpha\eta}\mathrm{d}\eta$$

and α must be a solution of this equation. For convergence of the integral, all possible α must lie in the domain

$$-\tfrac{1}{2} < \operatorname{Re}\alpha < \tfrac{1}{2}$$

Using Fourier transforms, it follows that

$$1 = \pi\lambda\sec\pi\alpha$$

Thus if α is a root, so also is $-\alpha$.

It will be noted that if $\lambda = \pi^{-1}$, there is a double root at $\alpha = 0$, and it will be found in fact that the equation for $\psi(\xi)$ is satisfied by $\psi(\xi) = 1$ and $\psi(\xi) = \xi$. Thus

$$\begin{aligned}\phi(x) &= \mathrm{e}^{-\frac{1}{2}\xi}\psi(\xi) \\ &= x^{\alpha-\frac{1}{2}} \quad \text{where} \quad \alpha = \frac{1}{\pi}\sec^{-1}(\pi\lambda) \\ &= x^{-\frac{1}{2}}(A + B\log x) \quad \text{if} \quad \lambda\pi = 1\end{aligned}$$

Example 4.8

Solve the integral equation for $\phi(t)$

$$f(x) = \int_{-\infty}^{\infty} k(x-t)\phi(t)\mathrm{d}t$$

Exactly as in Example 4.6, it can be shown that the principal solution is

$$\phi(x) = \frac{1}{2\pi}\int_{-\infty}^{\infty} \frac{F(\omega)}{K(\omega)} e^{-i\omega x} d\omega$$

provided that the operations are valid. It can be seen that there is a complementary function $e^{\alpha x}$ if,

$$0 = \int_{-\infty}^{\infty} k(u) e^{-\alpha u} du$$

Example 4.9

Solve the Stieltje's integral equation

$$f(x) = \int_0^{\infty} \frac{g(y) dy}{x+y}$$

Let

$$x = e^{\xi}, \quad y = e^{\eta}, \quad e^{\frac{1}{2}\xi} f(x) = p(\xi), \quad e^{\frac{1}{2}\eta} g(y) = q(\eta)$$

The integral equation becomes

$$p(\xi) = \int_{-\infty}^{\infty} \frac{q(\eta) d\eta}{2 \cosh \frac{1}{2}(\xi - \eta)}$$

and from Example 4.8, it follows by complex integration that

$$q(\xi) = \frac{1}{2\pi}[p(\xi + i\pi) + p(\xi - i\pi)]$$

$$g(x) = \frac{i}{2\pi}[f(x e^{i\pi}) - f(x e^{-i\pi})]$$

It will be noted that the terms here will have to be interpreted in the sense of complex variable theory.

Example 4.10 (Stieltje's Moment Problem)

Determine $f(x)$ such that

$$\int_0^{\infty} x^n f(x) dx = c_n \qquad (n = 0, 1, \ldots)$$

The sequence $\{c_n\}$ is some given sequence of numbers with the property that the series

$$C(u) = \sum_{n=1}^{\infty} \frac{(-1)^n c_n u^{2n}}{(2n)!}$$

is convergent.

$$C(u) = \int_0^\infty f(x) \sum_{n=0}^\infty \frac{(-1)^n x^n u^{2n}}{(2n)!} \mathrm{d}x$$
$$= \int_0^\infty f(x) \cos u\sqrt{x}\,\mathrm{d}x$$
$$= 2\int_0^\infty y f(y^2) \cos uy\,\mathrm{d}y$$

The principal solution of this is given by

$$y f(y^2) = \frac{1}{\pi}\int_0^\infty C(u) \cos uy\,\mathrm{d}u$$

The principal solutions may in fact be generalized functions, rather than functions if the c_n are not of a suitable form, e.g. if

$$c_n = 1, \quad C(u) = \cos u$$

and formally

$$2y f(y^2) = \delta(y-1)$$

It may be noted that the solution is not unique and that complementary functions exist. This is so because

$$\int_0^\infty x^n \mathrm{e}^{-x^\mu \cos\alpha} \sin(x^\mu \sin\alpha)\,\mathrm{d}x = \mu^{-1}\Gamma\left(\frac{n+1}{\mu}\right)\sin\left(\frac{n+1}{\mu}\right)\alpha$$

if $\mu > 0$ and $0 < \alpha < \dfrac{\pi}{2}$

Putting $\alpha = \mu\pi$, n an integer

$$\int_0^\infty x^n \mathrm{e}^{-x^\mu \cos\mu\pi} \sin(x^\mu \sin\mu\pi)\mathrm{d}x = 0$$

for all μ such that $0 < \mu < \frac{1}{2}$ and so $\mathrm{e}^{-x^\mu \cos\mu\pi}\sin(x^\mu \sin\mu\pi)$ is a complementary function for $0 < \mu < \frac{1}{2}$.

4.3 Laplace Integral Equations

A discussion has already been given in Section 3.3 of some integral equations (convolution Volterra equations) which can be solved with the aid of the Laplace transform. There are a number of other integral equations which can also be solved by the use of the Laplace transform theory. The most obvious application is the use of the inverse transform.

If $f(t)$ is defined by the integral equation

$$F(p) = \int_0^\infty \mathrm{e}^{-pt} f(t)\mathrm{d}t \qquad (4.7)$$

then

$$f(t) = \frac{1}{2\pi i} \int_{c-i\infty}^{c+i\infty} e^{pt} F(p) dp \tag{4.8}$$

where c is some real number which is greater than the real part of all the poles of $F(p)$. Equation (4.7) represents the Laplace transformation and Eq. (4.8) represents the inverse Laplace transformation. Much has been written about this inversion process so only one elementary example will be given. It will be seen that $f(t)$ will not be defined for negative t because the integral on the right-hand side of Eq. (4.7) is over $[0, \infty]$ and the values of f for negative t are not involved in the posing of the problem.

Example 4.11

Solve the integral equation

$$\frac{a}{a^2+p^2} = \int_0^\infty e^{-pt} f(t) dt \qquad a > 0$$

$$F(p) = \frac{a}{a^2+p^2} = \frac{1}{2i}\left\{\frac{1}{p-ai} + \frac{1}{p+ai}\right\}$$

Take $c > a$. Then

$$f(t) = \frac{1}{2\pi i} \int_{c-i\infty}^{c+i\infty} e^{pt} \left\{\frac{1}{2i}\left(\frac{1}{p-ai} + \frac{1}{p+ai}\right)\right\} dp$$

By the methods of the complex integral calculus, it follows that

$$f(t) = \frac{1}{2i}(e^{iat} - e^{-iat}) = \sin at \qquad t > 0$$

4.4 Hilbert Transform

If

$$g(u) = \frac{1}{\pi} \int_{-\infty}^{*\infty} \frac{f(x) dx}{x-u} \tag{4.9}$$

then

$$f(x) = \frac{1}{\pi} \int_{-\infty}^{*\infty} \frac{g(u) du}{x-u} \tag{4.10}$$

g is said to be the Hilbert transform of f, and f the inverse Hilbert transform of g. The solution of Eq. (4.10) is given by Eq. (4.9) and vice versa. It may be remembered that solutions need not always be functions, but may be generalized functions. For the evaluation of the integral, it may be convenient to proceed in one of the following ways: consider Eq. (4.9)

$$g(u) = \frac{1}{\pi} \lim_{R\to\infty} \int_{-R}^{*R} \frac{f(x)\mathrm{d}x}{x-u}$$

$$= 2i \lim_{R\to\infty} \frac{1}{2\pi i} \int_{\mathscr{C}} \frac{f(z)\mathrm{d}z}{z-u} \tag{4.11}$$

where $\mathscr{C}$ is the contour in the z plane consisting of the portion of the real axis from $x = -R$ to $x = +R$ and the semicircle of radius R, centre the origin, taken in the anticlockwise direction. The condition for this last step to hold is that $f(z) = 0(1)$ for large R. Using the methods of the complex integral calculus, it follows that

$$g(u) = 2i\left[\tfrac{1}{2}f(u) + \tfrac{1}{2}\sum_{\alpha} r(x_\alpha)(x_\alpha - u)^{-1} + \sum_{\beta} r(z_\beta)(z_\beta - u)^{-1}\right] \tag{4.12}$$

$r(x_\alpha)$ is the residue of $f(z)$ at a pole x_α on the real axis and $r(z_\beta)$ is the residue of $f(z)$ at a pole z_β in the upper half plane. Expression (4.11) can also be written in the alternative form

$$\frac{1}{\pi} \lim_{R\to\infty} \int_{-R}^{*R} \frac{f(x) - f(u) + f(u)}{x-u}\mathrm{d}x$$

and using the results of Appendix A, this becomes

$$\frac{1}{\pi} \lim_{R\to\infty} \left[\int_{-R}^{R} \frac{f(x)-f(u)}{x-u}\mathrm{d}x + f(u)\log\left(\frac{R-u}{R+u}\right)\right] = \frac{1}{\pi}\int_{-\infty}^{\infty} \frac{f(x)-f(u)}{x-u}\mathrm{d}x \tag{4.13}$$

The integral becomes an ordinary integral because it is assumed that $[f(x)-f(u)]/(x-u)$ is finite near $x = u$.

A third method is the following. It is possible to write

$$\frac{1}{\pi}\int_{-\infty}^{*\infty} \frac{f(x)\mathrm{d}x}{x-u} = \frac{1}{2\pi}\int_{-\infty}^{*\infty} \left[\frac{f(u+x) - f(u-x) + f(u-x) + f(u+x)}{x}\right]\mathrm{d}x$$

The integral of the last two terms taken together is zero because $x^{-1}[f(u-x)+f(u+x)]$ is odd in x. It follows that

$$g(u) = \frac{1}{2\pi}\int_{-\infty}^{\infty} \frac{f(u+x)-f(u-x)}{x}\mathrm{d}x \tag{4.14}$$

The integral is now an ordinary integral because $f(u+x) - f(u-x)$ is zero at $x = 0$. This integral can be evaluated by the usual methods such as contour integration.

Example 4.12

Solve the integral equation

$$\frac{1}{a^2+x^2} = \frac{1}{\pi}\int_{-\infty}^{\infty} \frac{g(u)\mathrm{d}u}{x-u} \qquad (a > 0)$$

$$g(u) = \frac{1}{\pi}\int_{-\infty}^{\infty} \frac{\mathrm{d}x}{(a^2+x^2)(x-u)}$$

The integral could of course be evaluated by writing

$$\frac{1}{a^2+x^2}\,\frac{1}{x-u} = \frac{Ax+B}{a^2+x^2} + \frac{C}{x-u}$$

writing

$$\int_{-\infty}^{\infty} = \lim_{\varepsilon \to 0}\int_{-\infty}^{u-\varepsilon} + \int_{u+\varepsilon}^{\infty}$$

and integrating in the usual way. However, consider the methods already discussed.

The function $f(z) = 1/(z^2+a^2)$ has a pole at $z = ai$ in the upper half plane and no poles on the real axis. The first method gives

$$g(u) = i\left\{\frac{1}{a^2+u^2}\right\} + \frac{2i}{2ai}\,\frac{1}{(ai-u)}$$

$$= -\frac{u}{a^2+u^2}$$

The second method gives

$$g(u) = \frac{1}{\pi}\lim_{R\to\infty}\left[\int_{-R}^{R}\left(\frac{1}{a^2+x^2} - \frac{1}{a^2+u^2}\right)\frac{\mathrm{d}x}{x-u}\right]$$

$$= -\frac{1}{\pi}\frac{1}{(a^2+u^2)}\lim_{R\to\infty}\int_{-R}^{R}\frac{x+u}{a^2+x^2}\mathrm{d}x$$

$$= -\frac{1}{\pi}\frac{u}{(a^2+u^2)}\lim_{R\to\infty}\int_{-R}^{R}\frac{\mathrm{d}x}{a^2+x^2}$$

$$= -\frac{1}{a^2+u^2}$$

It will be noted that by taking the integral between finite limits, the non-convergence of the integral

$$\int_{-\infty}^{\infty}\frac{x}{a^2+x^2}\mathrm{d}x \text{ is immaterial}$$

The third method of integration is as follows:

$$g(u) = \frac{1}{2\pi}\int_{-\infty}^{\infty}\left[\frac{1}{a^2+(u+x)^2}-\frac{1}{a^2-(u-x)^2}\right]\frac{\mathrm{d}x}{x}$$

$$= -\frac{2u}{\pi}\int_{-\infty}^{\infty}\frac{\mathrm{d}x}{[a^2+(u+x)^2][a^2+(u-x)^2]}$$

$$= -4iu \lim_{R\to\infty}\frac{1}{2\pi i}\int_{\mathscr{C}}\frac{\mathrm{d}z}{[a^2+(u+z)^2][a^2+(u-z)^2]}$$

$\mathscr{C}$ is the contour in the z plane consisting of the portion of the real axis from $-R$ to R and the upper semicircle of radius R, centre the origin taken in the anticlockwise direction. The poles in the upper half plane are at

$$z = u+ai \quad \text{and} \quad z = -u+ai$$

It follows that

$$g(u) = -4iu\left[\frac{1}{2ai[(2u+ai)^2+a^2]}+\frac{1}{2ai[(2u-ai)^2+a^2]}\right] = -\frac{u}{a^2+u^2}$$

Example 4.13

Solve the integral equation

$$\frac{1}{a+x} = \frac{1}{\pi}\int_{-\infty}^{\infty}\frac{g(y)\,\mathrm{d}y}{x-y} \qquad \text{where } a \text{ is real}$$

It can be seen immediately that the formal solution is

$$g(y) = -\pi\delta(y+a)$$

4.5 Finite Hilbert Transforms

Consider the function $g(u)$ defined by

$$g(u) = \frac{1}{\pi}\int_a^{*b}\frac{f(x)}{x-u}\mathrm{d}x \qquad a \leqslant u \leqslant b$$

$g(u)$ is the finite Hilbert transform of $f(x)$.

It is clear that, by a change of variable, it is always possible to consider this in the more symmetrical form

$$g(u) = \frac{1}{\pi}\int_{-1}^{*+1}\frac{f(x)\,\mathrm{d}x}{x-u} \qquad -1 \leqslant u \leqslant 1 \tag{4.15}$$

Although there is a superficial simplicity between the Hilbert transform and the finite Hilbert transform, the method of inversion is quite different.

Let

$$x = \cos\phi, \quad u = \cos\theta$$

With this change of variables, Eq. (4.15) becomes

$$g(\cos\theta) = \frac{1}{\pi}\int_0^{\pi} \frac{f(\cos\phi)\sin\phi\,\mathrm{d}\phi}{\cos\phi - \cos\theta} \qquad 0 \leqslant \theta \leqslant \pi \tag{4.16}$$

and it is this form of the transform that gives the hint to the method of solution.

As a preliminary, it is necessary to note that

$$I_n = \frac{1}{\pi}\int_0^{*\pi} \frac{\cos n\phi\,\mathrm{d}\phi}{\cos\phi - \cos\theta} = \frac{1}{2\pi}\int_{-\pi}^{*\pi} \frac{\cos n\phi\,\mathrm{d}\phi}{\cos\phi - \cos\theta}$$
$$= \sin n\theta \operatorname{cosec}\theta \text{ if } n \text{ is an integer}$$

This follows by putting $\zeta = \mathrm{e}^{i\phi}$ and integrating around the circle $|\zeta| = 1$.

It follows, by considering $I_{n+1} - I_{n-1}$ that

$$\frac{1}{\pi}\int_0^{*\pi} \frac{\sin n\phi \sin\phi\,\mathrm{d}\phi}{\cos\phi - \cos\theta} = -\cos n\theta$$

It may be noted also that it is possible to express any piecewise continuous function as a half range series in either series or cosines over $(0, \pi)$.

Let

$$F(\phi) = \tfrac{1}{2}a_0 + \sum_{n=1}^{\infty} a_n \cos n\phi$$

Then

$$\frac{1}{\pi}\int_0^{*\pi} \frac{F(\phi)\,\mathrm{d}\phi}{\cos\phi - \cos\theta} = \sum_{n=1}^{\infty} a_n \sin n\theta \operatorname{cosec}\theta$$

Thus, if

$$G(\theta) = \sum_{n=1}^{\infty} b_n \sin n\theta \tag{4.17}$$

ie

$$b_n = \frac{2}{\pi}\int_0^{\pi} G(\theta)\sin n\theta\,\mathrm{d}\theta$$

it follows that the solution of the integral equation

$$\frac{1}{\pi}\int_0^{*\pi} \frac{F(\phi)\,\mathrm{d}\phi}{\cos\phi - \cos\theta} = G(\theta)\operatorname{cosec}\theta \tag{4.18}$$

is given by

$$F(\phi) = \tfrac{1}{2}b_0 + \sum_{n=1}^{\infty} b_n \sin n\phi \tag{4.19}$$

This follows because, equating coefficients, $a_n = b_n$, $n > 0$; b_0 is arbitrary.
Also

$$\begin{aligned}\frac{1}{\pi}\int_0^{*\pi} \frac{G(\theta)\sin\theta\,\mathrm{d}\theta}{\cos\theta - \cos\phi} &= \frac{1}{\pi}\int_0^{*\pi} \frac{\sum_{n=1}^{\infty} b_n \sin n\theta \sin\theta\,\mathrm{d}\theta}{\cos\theta - \cos\phi} \\ &= -\sum_{n=1}^{\infty} b_n \cos n\phi \\ &= \tfrac{1}{2}b_0 - F(\phi) = \frac{1}{\pi}\int_0^{\pi} F(\phi)\,\mathrm{d}\phi - F(\phi)\end{aligned}$$

Thus

$$F(\phi) = \frac{1}{\pi}\int_0^{\pi} F(\phi)\,\mathrm{d}\phi - \frac{1}{\pi}\int_0^{*\pi} \frac{G(\theta)\sin\theta\,\mathrm{d}\theta}{\cos\theta - \cos\phi} \tag{4.20}$$

The first term on the right hand side of Eq. (4.20) is associated with the fact that F involves an arbitrary constant which is virtually its mean value over the range of definition. It will be seen that the solution of Eq. (4.18) for F is given by Eq. (4.20) and the solution of Eq. (4.20) for G by Eq. (4.18). The solution of Eq. (4.15) follows. Equations (4.16) and (4.18) are identical if

$$F(\phi) = F(\cos^{-1} x) = \sqrt{(1-x^2)}f(x) \tag{4.21}$$

$$G(\theta) = G(\cos^{-1} u) = \sqrt{(1-u^2)}g(u) \tag{4.22}$$

The solution of Eq. (4.16) is therefore given, from Eq. (4.20), by

$$\begin{aligned}\sqrt{(1-x^2)}f(x) &= \frac{1}{\pi}\int_0^{\pi} F(\phi)\,\mathrm{d}\phi - \frac{1}{\pi}\int_0^{*\pi} \frac{\sqrt{(1-u^2)}g(u)\sqrt{(1-u^2)}\,\mathrm{d}\phi}{\cos\phi - \cos\theta} \\ &= \frac{1}{\pi}\int_{-1}^{1} f(x)\,\mathrm{d}x + \frac{1}{\pi}\int_{-1}^{*1} \frac{(1-u^2)^{\frac{1}{2}}g(u)\,\mathrm{d}u}{(u-x)}\end{aligned} \tag{4.23}$$

Because of the previous working

$$\int_0^{\pi} F(\phi)\,\mathrm{d}\phi = \int_{-1}^{1} f(x)\,\mathrm{d}x \text{ is arbitrary} = \pi C\text{, say}$$

Thus

$$f(x) = (1-x^2)^{-\frac{1}{2}}\left\{C + \frac{1}{\pi}\int_{-1}^{*1}\frac{(1-u^2)^{\frac{1}{2}}g(u)\,\mathrm{d}u}{u-x}\right\} \tag{4.24}$$

and this is the solution of the integral equation (4.15). The solution is therefore not unique, and some other condition such as $f(x_0) = f_0$ must be imposed. However it is not so simple at the end points. Because of the factor $(1-x^2)^{-\frac{1}{2}}$, the condition there is that f is finite. Consider therefore the condition that $f(-1)$ is finite. Then, it follows from Eq. (4.23) that

$$\int_{-1}^{1} f(x)\,\mathrm{d}x + \int_{-1}^{*1}\frac{(1-u^2)^{\frac{1}{2}}g(u)\,\mathrm{d}u}{1+u} = 0$$

whence

$$(1-x^2)^{\frac{1}{2}} f(x) = \frac{1}{\pi}\int_{-1}^{*1}(1-u^2)^{\frac{1}{2}}g(u)\left(\frac{1}{u-x} - \frac{1}{u+1}\right)\mathrm{d}u$$

$$= \frac{1}{\pi}\int_{-1}^{*1}(1-u^2)^{\frac{1}{2}}g(u)\frac{(1+x)}{(u-x)(u+1)}\mathrm{d}u$$

whence

$$f(x) = \frac{1}{\pi}\left(\frac{1+x}{1-x}\right)^{\frac{1}{2}}\int_{-1}^{*l}\left(\frac{1-u}{1+u}\right)^{\frac{1}{2}}\frac{g(u)\,\mathrm{d}u}{u-x} \tag{4.25}$$

Example 4.13

Solve the integral equation

$$\int_{0}^{*l}\frac{h(u)\,\mathrm{d}u}{u-\omega} = 1 \qquad 0 \leqslant \omega \leqslant l$$

Let

$$u = \frac{l}{2}(\xi+1), \quad \omega = \frac{l}{2}(\eta+1)$$

Let

$$h(u) = f(\xi)$$

The integral equation becomes

$$\frac{1}{\pi}\int_{-1}^{*1}\frac{f(\xi)\,\mathrm{d}\xi}{\xi-\eta} = \frac{2}{\pi} \qquad -1 \leqslant \eta \leqslant 1$$

Let

$$\xi = \cos\phi, \quad \eta = \cos\theta, \quad \phi = \cos^{-1}\left(\frac{2u}{l} - 1\right)$$

The integral equation becomes

$$\frac{1}{\pi}\int_0^{*\pi} \frac{f(\cos\phi)\sin\phi \,\mathrm{d}\phi}{\cos\phi - \cos\theta} = \frac{2}{\pi} \tag{A}$$

Now

$$\frac{1}{\pi}\int_0^{*\pi} \frac{\mathrm{d}\phi}{\cos\phi - \cos\theta} = 1$$

$$\frac{1}{\pi}\int_0^{*\pi} \frac{\mathrm{d}\phi}{\cos\phi - \cos\theta} = 0$$

It follows that the integral equation (A) is satisfied by

$$f(\cos\phi)\sin\phi = \frac{2\cos\phi}{\pi} + K$$

where K is arbitrary. The result follows immediately.

Example 4.14

Solve the integral equation

$$\int_{-\pi}^{\pi} \phi(y)\log|\cos x - \cos y|\mathrm{d}y = f(x) \qquad -\pi \leqslant x \leqslant \pi$$

Two statements may be made immediately. $f(x)$ must be even in x or the two sides of the equation are not consistent, and there is a complementary function which can be any odd function of y. If ϕ is the solution of the equation which is even in y, it follows that

$$\int_0^{\pi} \phi(y)\log|\cos x - \cos y|\mathrm{d}y = \tfrac{1}{2}f(x) \qquad 0 \leqslant x \leqslant \pi$$

Let

$$\cos x = \xi, \quad \cos y = \eta$$

The integral equation becomes

$$\int_{-1}^{+1} \phi(\cos^{-1}\eta)\log|\xi - \eta|\frac{\mathrm{d}\eta}{\sqrt{(1-\eta^2)}} = \tfrac{1}{2}f(\cos^{-1}\xi) \qquad -1 \leqslant \xi \leqslant 1$$

Let

$$\phi(\cos^{-1}\eta) = \sqrt{(1-\eta^2)}\Phi(\eta)$$

$$\tfrac{1}{2}f(\cos^{-1}\xi) = F(\xi)$$

The integral equation becomes

$$\int_{-1}^{+1} \Phi(\eta) \log|\xi-\eta| \mathrm{d}\eta = F(\xi)$$

Differentiating with respect to ξ, it follows that

$$\int_{-1}^{*+1} \frac{\Phi(\eta)}{\xi-\eta} \mathrm{d}\eta = F'(\xi)$$

Thus

$$(1-\eta^2)^{\frac{1}{2}}\Phi(\eta) = C + \frac{1}{\pi} \int_{-1}^{*+1} \frac{(1-u^2)^{\frac{1}{2}}F'(u)\mathrm{d}u}{u-\eta}$$

and

$$\phi(y) = C + \frac{1}{2\pi} \int_{-1}^{*+1} \frac{f'(z)\mathrm{d}z}{\cos z - \cos y}$$

4.6 Miscellaneous Integral Transforms

There are of course many other integral transforms. A selection of these is given here, together with their inverses for reference, but no examples or exercises will be involved. For details of the transformations see Reference 6. In all cases $f(x)$ is the unknown function which it is desired to find.

(a) Mellin Transform

$$\bar{f}(s) = \int_0^\infty f(x) x^{s-1} \, \mathrm{d}x \tag{4.26}$$

$$f(x) = \frac{1}{2\pi i} \int_{c-i\infty}^{c+i\infty} \bar{f}(s) x^{-s} \, \mathrm{d}x \tag{4.27}$$

c is such that $\bar{f}(s)$ is defined everywhere on the contour and the integral exists.

(b) Hankel Transform

$$\bar{f}(\xi) = \int_0^\infty x f(x) J_\nu(\xi x) \, \mathrm{d}x \tag{4.28}$$

$$f(x) = \int_0^\infty \xi \bar{f}(\xi) J_\nu(x\xi) \, \mathrm{d}\xi \tag{4.29}$$

(c) MacRobert Transform

$$g(u) = \int_p^q xf(x)J_x(xu)\mathrm{d}x \qquad 0 \leqslant p < q \tag{4.30}$$

$$f(x) = \int_0^\infty (u-u^{-1})g(u)J_x(xu)\mathrm{d}u \qquad p < x < q \tag{4.31}$$

(d) Legendre Transform

Let

$$\bar{f}_L(n) = \int_{-1}^{+1} f(x)P_n(x)\mathrm{d}x \tag{4.32}$$

Clearly, if n is even, any odd function is a complementary function and if n is odd, any even function is a complementary function. For n even

$$f(x) = \frac{1}{2}\sum_{n=0}^{\infty}(4n+1)\bar{f}_L(2n)P_{2n}(x) \tag{4.33}$$

and for n odd

$$f(x) = \frac{1}{2}\sum_{n=1}^{\infty}(4n-1)\bar{f}_L(2n-1)P_{2n-1}(x) \tag{4.34}$$

In all cases, $f(x)$ is to be replaced whenever necessary by

$$\tfrac{1}{2}[f(x-0)+f(x+0)]$$

EXERCISES

1. Solve the integral equation

$$\frac{x}{a^2+x^2} = \int_0^\infty \sin \omega x\phi(\omega)\mathrm{d}\omega \qquad a > 0$$

2. Find the eigenvalues and eigenfunctions of the integral equation

$$\phi(x) = \lambda\int_0^\infty \sin xy\phi(y)\,\mathrm{d}y$$

3. Find the solution of the integral equation

$$\phi(x) = \mathrm{e}^{-ax} + \lambda\int_0^\infty \sin xy\phi(y)\mathrm{d}y$$

$$a > 0 \qquad (\pi\lambda^2 \neq 2)$$

4. Solve the integral equation

$$\frac{p}{p^2+a^2} = \int_0^\infty e^{-pt} f(t) dt \qquad a > 0$$

5. Find the principal solution of

$$\phi(x) = e^{-a|x|} + \lambda \int_0^\infty e^{by} \phi(x-y) dy$$

and show that the complementary function is $e^{\alpha x}$ where $\alpha = \lambda + b$ if $\lambda > 0$.

6. Prove that the principal solution of

$$\phi(x) = f(x) + \int_{-\infty}^{\infty} k(x+y)\phi(y) dy$$

is given by

$$\phi(x) = \frac{1}{\sqrt{(2\pi)}} \int_{-\infty}^{\infty} F(\omega) e^{-i\omega x} d\omega$$

where $[1-2\pi K(\omega)K(-\omega)]F(\omega) = F(\omega) + \sqrt{(2\pi)} F(-\omega) K(\omega)$ provided that the operations are legitimate (Fox's integral equation).

7. Show that a solution of the form $e^{\alpha x}$ exists to the integro-differential equation

$$f'(x) = \lambda \int_0^\infty e^{-y} \left[\frac{f(x+y) - f(x-y)}{y} \right] dy$$

provided that

$$\alpha + \frac{\lambda}{2} \log\left(\frac{1+\alpha}{1-\alpha}\right) = 0 \quad \text{and} \quad |\alpha| < 1$$

8. Prove that the integral equation

$$g(x) = \frac{1}{\pi} \int_{-\infty}^{\infty} \frac{\sin a(x-y)}{x-y} \phi(y) dy \qquad (a > 0)$$

is satisfied by the principal solution

$$\phi(x) = x \int_{-a}^{a} g(\omega) e^{-ix\omega} d\omega$$

and show that 1 is a complementary function.

9. Determine $f(x)$ such that

(a) $$\int_0^\infty x^{2n} f(x)\mathrm{d}x = \mathrm{d}_n \qquad (n = 0, 1, \ldots)$$

(b) $$\int_0^\infty x^{2n-1} f(x)\mathrm{d}x = \mathrm{e}_n \qquad (n = 1, 2, \ldots)$$

10. Solve the integral equation

$$\frac{1}{(x^2+a^2)(x^2+b^2)} = \frac{1}{\pi} \int_{-\infty}^{*\infty} \frac{g(y)\mathrm{d}y}{x-y} \qquad (a, b > 0)$$

11. Solve the integral equation

$$\frac{1}{(x+a)^2} = \frac{1}{\pi} \int_{-\infty}^{*\infty} \frac{g(y)\mathrm{d}y}{x-y} \qquad (a \text{ real})$$

12. Solve the integral equation

$$\int_{-1}^{*1} \frac{f(x)\mathrm{d}x}{x-y} = y^2$$

13. Prove that the solution of the integral equation

$$Q(x) = \frac{\mathrm{d}}{\mathrm{d}x} \int_1^x \frac{P(y)\mathrm{d}y}{(y^2-x^2)^{\frac{1}{2}}}$$

is

$$P(y) = 2y \int_y^1 \frac{Q(z)\mathrm{d}z}{(z^2-y^2)^{\frac{1}{2}}}$$

14. Prove that the solution of

$$\psi(x) = \pi \int_x^\infty \frac{f'(u)\mathrm{d}u}{(u^2-x^2)^{\frac{1}{2}}}$$

is

$$f(u) = -2 \int_u^\infty \frac{y\psi(y)\mathrm{d}y}{(y^2-u^2)^{\frac{1}{2}}} + \text{const}$$

5 Approximate Methods

5.1 General

In this chapter certain elementary aspects of the solution of non-linear integral equations and certain approximate methods associated with the solution of integral equations will be considered. To a very great extent, the treatment of non-linear equations will be *ad hoc* in nature as there are no real rules by which non-linear equations can be solved, apart from the possibility of iteration. The methods suggested may work only with the actual example indicated, but the approach may give an idea of some possible way of attacking another problem. It will be noted that the majority of worked examples in this chapter are fairly simple and that, for example, integrations may be represented merely by two term approximation, to ensure that the principles behind the method do not get lost in a mass of arithmetical calculations and algebraic manipulations.

5.2 Non-linear Volterra Equations

Consider the non-linear Volterra equation of the second kind

$$\phi(x) = f(x)+\lambda \int_0^x F\{x, y, \phi(y)\}\,\mathrm{d}y \tag{5.1}$$

(By a change of variable any other value could replace zero as lower limit.) Formally, it is possible to construct a sequence of functions $\phi^{(n)}(x)$ defined by

$$\phi^{(n)}(x) = f(x)+\lambda \int_0^x F\{x, y, \phi^{(n-1)}(y)\}\,\mathrm{d}y \qquad n \geqslant 1 \tag{5.2a}$$

$$\phi^{(0)}(x) = f(x) \tag{5.2b}$$

The question arises: under what conditions will this sequence of functions converge to the solution of Eq. (5.1)? The following assumptions are made: They may not be necessary for the convergence of the sequence to the solution, but it will be useful to obtain sufficient conditions for convergence. The reader who has studied differential equations, will recognize, in the method discussed below, Picard's method. A short discussion appears in Section 3.4(b).

Suppose that $f(x)$ is continuous in $0 \leqslant x \leqslant X$ (X is finite but unspecified), that $F(x, y, z)$ is continuous with respect to all variables in $0 \leqslant x \leqslant X$, $0 \leqslant y \leqslant x$, $\alpha \leqslant z \leqslant \beta$ that $\alpha < f(x) < \beta$ and that $|f(x)| < \gamma$. Furthermore $F(x, y, z)$ satisfies the Lipschitz condition

$$|F(x,y,z') - F(x,y,z'')| < k|z' - z''|$$

k being a positive constant, and $|F(x,y,z)| < M$. Now

$$\phi^{(n)}(x) = \sum_{s=1}^{n} \{\phi^{(s)}(x) - \phi^{(s-1)}(x)\} + \phi^{(0)}(x)$$

and so the convergence of the sequence $\{\phi^{(n)}(x)\}$ is equivalent to the convergence of the series whose sth term is

$$\phi^{(s)}(x) - \phi^{(s-1)}(x)$$

Now

$$|\phi^{(s)}(x) - \phi^{(s-1)}(x)| = |\lambda| \left| \int_0^x F\{x, y, \phi^{(s-1)}(y)\} - F\{x, y, \phi^{(s-2)}(y)\}\, \mathrm{d}y \right|$$

$$\leqslant |\lambda| \int_0^x k|\phi^{(s-1)}(y) - \phi^{(s-2)}(y)|\, \mathrm{d}y \qquad (5.3)$$

Also

$$\phi^{(1)}(x) - \phi^{(0)}(x) = \lambda \int_0^x F\{x, y, f(y)\}\, \mathrm{d}y$$

as

$$\phi^{(0)}(x) = f(x)$$

Then

$$|\phi^{(1)}(x) - \phi^{(0)}(x)| \leqslant |\lambda| \int_0^x |F\{x, y, f(y)\}|\, \mathrm{d}y \leqslant |\lambda| \int_0^x M\, \mathrm{d}y = M|\lambda|x$$

By induction it follows that

$$|\phi^{(s)}(x) - \phi^{(s-1)}(x)| \leqslant \frac{(k|\lambda|x)^s}{s!} M$$

This is the sth term of the power series for $M \exp\{k|\lambda|x\}$, and so the series

$$\phi^{(0)}(x) + \sum_{s=1}^{\infty} \{\phi^{(s)}(x) - \phi^{(s-1)}(x)\}$$

is always absolutely and uniformly convergent no matter what the value of λ, and its sum

$$\lim_{n\to\infty} \phi^{(n)}(x)$$

will be the solution of the integral Eq. (5.1).

The error in stopping at n terms in the series is given by

$$R^{(n)}(x) = \phi(x) - \phi^{(n)}(x) = \sum_{s=n+1}^{\infty} \{\phi^{(s)}(x) - \phi^{(s-1)}(x)\}$$

and

$$|R^{(n)}(x)| \leqslant \sum_{s=n+1}^{\infty} |\phi^{(s)}(x) - \phi^{(s-1)}(x)| \leqslant M \sum_{s=n+1}^{\infty} \frac{(k|\lambda|x)^s}{s!}$$

$$= M\left\{\exp(k|\lambda|x) - \sum_{s=0}^{n} \frac{(k|\lambda|x)^s}{s!}\right\} \tag{5.4}$$

The uniqueness of the solution can be proved as follows: Suppose that $\phi(x)$ and $\psi(x)$ are two possible solutions. Then

$$\chi(x) = \phi(x) - \psi(x) = \lambda \int_0^x [F\{x, y, \phi(y)\} - F\{x, y, \psi(y)\}]\,\mathrm{d}y$$

and so by the Lipschitz condition

$$|\chi(x)| \leqslant |\lambda| k \int_0^x |\chi(y)|\,\mathrm{d}y$$

Let $\chi_{\max}$ be the maximum value of $\chi(x)$ in $0 \leqslant x \leqslant X$. Then

$$\chi_{\max} \leqslant |\lambda| k X \chi_{\max}$$

X is arbitrary and unspecified. If $X > \{|\lambda|k\}^{-1}$ the conditions on F will certainly hold in the smaller interval $0 < x < X^* < \{|\lambda|k\}^{-1}$ and so X can be replaced by X^* if necessary, and it is possible to assume that $|\lambda k X = \Gamma < 1$. Hence

$$\chi_{\max} \leqslant \Gamma \chi_{\max}$$

This is possible only if $\chi_{\max}$ is zero, and so the solution is unique.

A non-linear Volterra equation of the first kind

$$\int_0^x F\{x, y, \phi(y)\}\,\mathrm{d}y = f(x) \tag{5.5}$$

can be dealt with in the following way: Differentiation with respect to x gives

$$F\{x, x, \phi(x)\} + \int_0^x \frac{\partial F}{\partial x}\{x, y, \phi(y)\}\,\mathrm{d}y = f'(x)$$

A second differentiation gives

$$\left[2\frac{\partial F}{\partial x}(x,y,z)+\frac{\partial F}{\partial y}(x,y,z)\right]+\left[\frac{\partial F}{\partial z}(x,y,z)\right]\phi'(x)$$
$$+\int_0^x \frac{\partial^2 F}{\partial x^2}\{x,y,\phi(y)\}\,\mathrm{d}y=f''(x) \quad (5.6)$$

The quantities in square brackets are to be evaluated at $y = x$ and $z = \phi(x)$. A sequence can be set up in the form

$$\phi^{(n)'}(x) = G\{\phi^{(n-1)}(x)\}$$

where

$$\phi^{(n)}(0) = \xi$$

and ξ is defined by

$$F(0,0,\xi) = f'(0)$$

The convergence conditions are too complicated to be indicated here. It may of course be possible for the solution to appear directly from the differentiation.

It may be shown that, if the conditions specified previously as applying to F are not satisfied, then there is no solution to the integral equation of the second kind that does not contain a free term $f(x)$.

$$\phi(x) = \lambda\int_0^x F\{x,y,\phi(y)\}\,\mathrm{d}y \quad (5.7)$$

provided that $F(x,y,0)$ is zero.

For the first term in the iterative sequence

$$\phi_1(x) = f(x)+\lambda\int_0^x F\{x,y,\phi(y)\}\,\mathrm{d}y = 0 \qquad \text{as} \qquad f(x)=0$$

It follows that all the other members of the sequence $\phi_n(x)$ will be zero and that the limit of the sequence, which is the solution of the equation, will be zero.

Example 5.1

Solve the integral equation

$$\phi(x) = \lambda\int_0^x \{1+\phi(y)\}^2\,\mathrm{d}y$$

Differentiating with respect to x

$$\phi'(x) = \lambda\{1+\phi(x)\}^2 \qquad \text{also} \qquad \phi(0)=0$$

The solution of this is clearly $\tan \lambda x$. In this case there is a solution because

$$F(x, y, z) = 1+z^2 \qquad \text{and} \qquad F(x, y, 0) \neq 0$$

Example 5.2

Find the first three functions in the sequence of functions arising from the iterative solution of the integral equation

$$\phi(x) = x+\lambda \int_0^x [1+x\{\phi(y)\}]^2 \, dy$$

$$\phi^{(0)}(x) = x$$

$$\phi^{(1)}(x) = x+\lambda \int_0^x (1+xy^2) \, dy$$

$$= (1+\lambda)x + \frac{\lambda x^4}{3}$$

$$\phi^{(2)}(x) = x+\lambda \int_0^x \left[1+x\left\{(1+\lambda)y + \frac{\lambda y^4}{3}\right\}^2 \right] dy$$

$$= (1+\lambda)x + \frac{\lambda x^4}{3}(1+\lambda) + \frac{\lambda^2}{9}(1+\lambda)x^7 + \frac{\lambda^3 x^{10}}{81}$$

5.3 Non-linear Fredholm Equations

A non-linear integral equation of the form

$$\phi(x) = f(x)+\lambda \int F\{x, y, \phi(y)\} \, dy \tag{5.8}$$

is termed an Urysohn equation.

A particular case is the Hammerstein equation

$$\phi(x) = f(x)+\lambda \int K(x, y) F\{y, \phi(y)\} \, dy \tag{5.9}$$

Clearly it will be better whenever possible to discuss the more general equation and this will be done. In this case again, an attempt will be made to construct a sequence of functions which will tend to a solution, and sufficient conditions will be given for the solution to be unique. The obvious sequence of functions to construct will be

$$\phi^{(0)}(x) = f(x) \tag{5.10a}$$

$$\phi^{(n)}(x) = \lambda \int F\{x, y, \phi^{(n-1)}(y)\} \, dy + f(x), \qquad n > 0 \tag{5.10b}$$

The conditions which will be imposed are as follows. It will be assumed for convenience that x and y lie in the domain $a \leqslant x, y \leqslant b$. It will be seen that suitable modifications can be made quite easily to the analysis for a problem involving more than one variable. Let $F(x, y, z)$ satisfy the Lipschitz condition

$$|F(x, y, z') - F(x, y, z'')| \leqslant k|z' - z''|$$

where k is a positive constant and let

$$|F(x, y, z)| < M$$

where M is a positive constant. Now

$$\phi^{(n)}(x) = \sum_{s=1}^{n} \{\phi^{(s)}(x) - \phi^{(s-1)}(x)\} + \phi_0(x)$$

and so the convergence of the sequence $\phi^{(n)}(x)$ is the same as that of the series $\phi^{(s)}(x) - \phi^{(s-1)}(x)$. Now

$$|\phi^{(s)}(x) - \phi^{(s-1)}(x)| = |\lambda| \left| \int \left\{ F\{x, y, \phi^{(s-1)}(y)\} - F\{x, y, \phi^{(s-2)}(y)\} \right\} \mathrm{d}y \right|$$

$$\leqslant |\lambda| \int |F\{x, y, \phi^{(s-1)}(y)\} - F\{x, y, \phi^{(s-2)}(y)\}| \, \mathrm{d}y$$

$$\leqslant |\lambda| k \int |\phi^{(s-1)}(y) - \phi^{(s-2)}(y)| \, \mathrm{d}y$$

also

$$\phi^{(1)}(x) - \phi^{(0)}(x) = \int F(x, y, f(y)) \, \mathrm{d}y = A(x)$$

It is assumed also that $A(x)$ is finite, $|A(x)| < L$. It follows that

$$|\phi^{(s)}(x) - \phi^{(s-1)}(x)| < [|\lambda| k(b-a)]^{s-1} L$$

Thus, if $|\lambda| k(b-a) < 1$ the series will be absolutely and uniformly convergent, and $\phi_n(x)$ will tend to some function $\phi(x)$ which will be the solution of Eq. (5.8).

It remains to determine the uniqueness of the solution. Let $\phi(x), \psi(x)$ be two solutions and let $\chi(x) = \phi(x) - \psi(x)$ and $\chi_{\max}$ be the maximum value of $\chi(x)$ in $a \leqslant x \leqslant b$. Then

$$\chi(x) = \lambda \int [F\{x, y, \phi(y)\} - F\{x, y, \psi(y)\}] \, \mathrm{d}y$$

Thus

$$|\chi(x)| \leqslant |\lambda| k \int \chi(y) \, \mathrm{d}y \leqslant |\lambda| k(b-a) \chi_{\max}$$

Thus

$$\chi_{\max} \leqslant |\lambda| k(b-a)\chi_{\max}$$

Now if

$$|\lambda| k(b-a) < 1, \qquad \chi_{\max} = 0$$

and so $\chi(x) = 0$. This is the condition that the iterative process converges.

Thus, if the iterative process converges, which is the case for λ small enough, it converges to a unique solution. This does not mean however that the solution is unique for arbitrary λ. The error in the approximation by stopping at the nth iteration is obtained as follows:

$$\begin{aligned}|R^{(n)}(x)| &= |\phi(x) - \phi^{(n)}(x)| \\ &= \left|\sum_{s=n+1}^{\infty} \phi^{(s)}(x) - \phi^{(s-1)}(x)\right| \\ &\leqslant \sum_{s=n+1}^{\infty} |\phi^{(s)}(x) - \phi^{(s-1)}(x)| \\ &< \sum_{s=n+1}^{\infty} \{|\lambda| k(b-a)\}^{s-1} A(x) \\ &= \frac{|\lambda| k(b-a)^n A(x)}{1 - |\lambda| k(b-a)}\end{aligned}$$

Before proceeding further, it should be noted that finding a solution for an equation of the first kind

$$\int F\{x, y, \phi(y)\}\, \mathrm{d}y = f(x) \tag{5.11}$$

is entirely a matter of luck, trial and error. There is no theory similar to the Hilbert Schmidt theory valid in this case. Consequently, the problem will not be discussed further, beyond remarking that it may be possible to find a convergent sequence of functions $\phi^{(n)}(x)$ defined by the relation

$$\phi^{(n+1)}(x) = \phi^{(n)}(x) + \mu\left[\int F\{x, y, \phi^{(n)}(y)\}\, \mathrm{d}y - f(x)\right]$$

where μ is some number. If the sequence converges, it will converge to a solution of the equation (5.11), but the solution may not be unique, and the solution obtained may depend upon the value of μ employed.

Consider now a non-linear Fredholm equation of the second kind without a free term. The equation

$$\phi(x) = \lambda \int F\{x, y, \phi(y)\}\, \mathrm{d}y \tag{5.12}$$

can have non-zero solutions which depend on λ, whereas in the linear case a non-singular kernel implies that non-zero solutions can exist only when λ is an eigenvalue, the associated functions being eigenvalues. Two remarks may be made about equations of the type Eq. (5.12). Firstly for a given λ there may be more than one solution to the integral equation, and secondly the solutions need not of necessity be real even though all the terms in the equation are formally real. It is possible for the solutions to be real for some values of real λ and complex for others. Generalizing previous ideas, solutions of Eq. (5.12) may be referred to as eigenfunctions corresponding to the eigenvalue λ. Re-writing Eq. (5.12) as

$$\int F\{x, y, \phi(y)\}\,\mathrm{d}y = \mu\phi(x) \tag{5.13}$$

it is possible to say that functions $\phi(y)$ such that

$$\int F\{x, y, \phi(y)\}\,\mathrm{d}y = 0$$

are eigenfunctions corresponding to $\mu = 0$. If now

$$\int F(x, y, 0)\,\mathrm{d}y = 0$$

then $\phi(x) = 0$ will be an eigenfunction irrespective of the value of λ, and in particular this will be so if $F(x, y, 0) = 0$.

Example 5.3

Find a first and second approximation in the iterative solution of the integral equation

$$\int_0^1 (x+y)^{\frac{1}{2}}[\phi(y)]^{\frac{1}{2}}\,\mathrm{d}y = \phi(x)$$

and finds bounds on $\phi(x)$.

The iterative sequence is

$$\phi^{(n)}(x) = \int_0^1 (x+y)^{\frac{1}{2}}[\phi^{(n-1)}(y)]^{\frac{1}{2}}\,\mathrm{d}y$$

Clearly it will not be any use starting with zero for the first approximation, as $\phi(x) = 0$ satisfies the equation identically. Consider therefore $\phi^{(1)}(x) = C$ where C is a constant which is determined by the relation

$$\int_0^1 C\,\mathrm{d}x = \int_0^1 \mathrm{d}x \int_0^1 (x+y)^{\frac{1}{2}} C^{\frac{1}{2}}\,\mathrm{d}y$$

as it is certainly true that

$$\int_0^1 \phi(x)\,dx = \int_0^1 \int_0^1 (x+y)^{\frac{1}{2}}[\phi(y)]^{\frac{1}{2}}\,dx\,dy$$

It can be shown that $C^{\frac{1}{2}} = \frac{4}{15}(2^{5/2}-1)$ and

$$\phi^{(2)}(x) = \int_0^1 (x+y)^{\frac{1}{2}} C^{\frac{1}{2}}\,dy = \tfrac{2}{3}C^{\frac{1}{2}}[(1+x)^{3/2} - x^{3/2}]$$

Bounds on $\phi(x)$ may be determined as follows [$\phi(x)$ must be positive]: Let

$$\phi_{max} = \max \phi(x) \qquad \text{in} \qquad 0 \leqslant x \leqslant 1$$

$$\phi_{min} = \min \phi(x) \qquad \text{in} \qquad 0 \leqslant x \leqslant 1$$

$$\phi_{max} \leqslant \int_0^1 (1+y)^{\frac{1}{2}} \phi_{max}^{\frac{1}{2}}\,dy = \tfrac{2}{3}(2^{3/2}-1)\,\phi_{max}^{\frac{1}{2}}$$

also

$$\phi(x) \leqslant \int_0^1 (x+y)^{\frac{1}{2}} \tfrac{2}{3}(2^{3/2}-1)\,dy = \tfrac{4}{9}(2^{3/2}-1)[(1+x)^{3/2} - x^{3/2}]$$

Similarly

$$\phi_{min} \geqslant \int_0^1 y^{\frac{1}{2}} \phi_{min}^{\frac{1}{2}}\,dy = \tfrac{2}{3}\phi_{min}^{\frac{1}{2}}$$

and so

$$\phi_{min}^{\frac{1}{2}} \geqslant \tfrac{2}{3}$$

It follows that

$$\phi(x) \geqslant \int_0^1 (x+y)^{\frac{1}{2}} \phi_{min}^{\frac{1}{2}}\,dy = \tfrac{4}{9}[(x+1)^{3/2} - x^{3/2}]$$

Thus lower and upper bounding functions have been found for $\phi(x)$.

5.4 Approximate Methods of Solution for Linear Integral Equations

There are a number of methods of finding approximate solutions for linear integral equations. In all cases, a set of functions is defined in some way, and the actual solution is a member of this set. An attempt is made to fit some member of the set of functions so that it has as many as possible, in some sense, of the properties of $\phi(x)$. If it had all, it would be $\phi(x)$, but it does not generally happen that one is able actually to light on the solution.

(a) Iterative Processes

Because linear integral equations can be treated as special cases of the integral equations discussed in Sections 5.1 and 5.2, it will be possible to apply the iterative processes developed there and to use the associated error estimating processes.

Example 5.4

Find the first three functions in the iterative solution of

$$\phi(x) = \lambda \int_0^1 \sin xy\phi(y)\,dy + 1$$

The first approximation is

$$\phi^{(0)}(x) = 1$$

The second is

$$\begin{aligned}\phi^{(1)}(x) &= \lambda \int_0^1 \sin xy\,dy + 1\\ &= \lambda x^{-1}(1-\cos x) + 1\end{aligned}$$

$$\begin{aligned}\phi^{(2)}(x) &= \lambda \int_0^1 \sin xy[1+\lambda y^{-1}(1-\cos y)]\,dy + 1\\ &= 1+\lambda x^{-1}(1-\cos x)+\lambda^2 \int_0^1 \sin xy(1-\cos y)y^{-1}\,dy\end{aligned}$$

Referring to Section 5.2, the quantities required for the error estimate are

$$A(x) = F\{x, y, f(y)\}\,dy = \int_0^1 \sin xy\,dy = x^{-1}(1-\cos x)$$

$$|F(x,y,z')-F(x,y,z'')| = |\sin xy(z'-z'')| \leqslant \sin 1|z'-z''|, \qquad k = \sin 1$$

$a = 0, b = 1$. Thus

$$|R^{(2)}(x)| \leqslant \frac{|\lambda|^2 \sin^2 1}{1-|\lambda|\sin 1}\left(\frac{1-\cos x}{x}\right)$$

provided that

$$|\lambda|\sin 1 < 1 \qquad \text{i.e.} \qquad |\lambda| < 1{\cdot}88$$

It is possible also to use an iterative process for a solution of an integral equation of the first kind

$$f(x) = \int K(x,y)\phi(y)\,dy$$

In this case, the kernel must be assumed non-singular and positive definite.

The kernel is then of the form

$$K(x, y) = \sum_{s=1}^{\infty} \frac{\phi_s(x)\phi_s(y)}{s}$$

where the ϕ_s are assumed to be a complete set of normalized eigenfunctions (see Appendix G), and the eigenvalues obey the relation

$$0 < \lambda_1 \leqslant \lambda_2 \ldots$$

If the set is not complete it will be necessary to add a complementary function to the solution.

Construct now a sequence of functions $\phi^{(n)}$ defined by

$$\phi^{(n)}(x) - \phi^{(n-1)}(x) = \lambda\left[f(x) - \int K(x, y)\phi^{(n-1)}(y)\,\mathrm{d}y\right] \tag{5.14}$$

If the sequence $\phi^{(n)}(x)$ converges, it will converge to the required solution. It remains to be seen under what conditions it will converge. Let

$$f(x) = \sum_{s=1}^{\infty} f_s\phi_s(x)$$

where

$$f_s = \int f(y)\phi_s(y)\,\mathrm{d}y$$

and

$$\phi(x) = \sum_{s=1}^{\infty} \lambda_s f_s \phi_s(x)$$

Equation (5.14) can be rewritten as

$$\sum_{s=1}^{\infty} c_s^{(n)}\phi_s(x) = \sum_{s=1}^{\infty} c_s^{(n-1)}\phi_s(x) + \lambda\left\{\sum_{s=1}^{\infty} f_s\phi_s(x) - \sum_{s=1}^{\infty} c_s^{(n-1)}\lambda_s^{-1}\phi_s(x)\right\}$$

Thus

$$c_s^{(n)} = c_s^{(n-1)}\left(1 - \frac{\lambda}{\lambda_s}\right) + \lambda f_s$$

and

$$c_s^{(n)} - c_s^{(n-1)} = \left(1 - \frac{\lambda}{\lambda_s}\right)(c_s^{(n-1)} - c_s^{(n-2)})$$

For the sequence $c_s^{(n)} - c_s^{(n-1)}$ to tend to zero (which is equivalent to $c_s^{(n)}$ tending to $\lambda_s f_s$), it follows that

$$-1 < 1 - \frac{\lambda}{\lambda_s} < 1 \qquad \text{i.e.} \qquad 0 < \lambda/\lambda_s < 2$$

Now $\lambda_s \geqslant \lambda_1 > 0$ and so the condition can be satisfied for all s only if $0 < \lambda < 2\lambda_1$. The reason why the kernel has to be positive definite now emerges. If the eigenvalues were not one signed, it would not be possible to satisfy $\lambda/\lambda_s > 0$ for all eigenvalues. Thus if $0 < \lambda < 2\lambda_1$ the sequence of functions $\phi^{(n)}(x)$ will converge to the solution of the integral equation. It may be remarked that, just as it is sometimes more convenient to solve a set of linear equations by means of an iterative process rather than direct elimination, it may be easier to solve an integral equation with a degenerate kernel by this iterative process than by the more orthodox method outlined previously. Some estimate of the error in the iteration process may be obtained as follows. Let

$$R^{(n)}(x) = \phi(x) - \phi^{(n)}(x)$$

Then

$$R^{(n)}(x) = R^{(n-1)}(x) - \lambda \int K(x, y) R^{n-1}(y)\,\mathrm{d}y$$

and if

$$R^{(n)}(x) = \sum_{s=1}^{\infty} r_s^{(n)} \phi_s(x)$$

$$r_s^{(n)} = r_s^{(n-1)}\left(1 - \frac{\lambda}{\lambda_s}\right)$$

and

$$r_s^{(n)} = \left(1 - \frac{\lambda}{\lambda_s}\right)^n r_s^{(0)}$$

The initial function in the iteration $\phi^{(0)}(x)$ has not yet been specified. If

$$\phi^{(0)}(x) = 0, \qquad R^{(0)}(x) = \phi(x) \qquad \text{and} \qquad r_s^{(0)} = \lambda_s f_s$$

and

$$r_s^{(n)} = \left(1 - \frac{\lambda}{\lambda_s}\right)^n \lambda_s f_s$$

Thus

$$\begin{aligned}
\int \{R^{(n)}(x)\}^2\,\mathrm{d}x &= \sum_{s=1}^{\infty} r_s^{(n)2} = \sum_{s=1}^{\infty} \lambda_s^2 f_s^2 (1 - \lambda/\lambda_s)^{2n} \\
&= \sum_{s=1}^{N} \lambda_s^2 f_s^2 (1 - \lambda/\lambda_s)^{2n} + \sum_{s=N+1}^{\infty} \lambda_s^2 f_s^2 (1 - \lambda/\lambda_s)^{2n} \\
&\leqslant \lambda_1^2 (1 - \lambda/\lambda_N)^{2n} \sum_{s=1}^{N} f_s^2 + \lambda_{N+1}^2 \sum_{s=N+1}^{\infty} f_s^2
\end{aligned}$$

as $\{\lambda_s\}$ forms a monotonic increasing sequence. This is true for any N. In particular

$$\int \{R^{(n)}(x)\}^2\,\mathrm{d}x \leqslant \lambda_1^2 \sum_{s=1}^{\infty} f_s^2 = \lambda_1^2 \int \{f(x)\}^2\,\mathrm{d}x$$

If $K(x, y)$ is a degenerate kernel with N terms,

$$\int \{R^{(n)}(x)\}^2 \, dx \leqslant \lambda_1^2 (1-\lambda/\lambda_N)^{2n} \sum_{s=1}^{N} f_s^2$$

$$= \lambda_1^2 (1-\lambda/\lambda_N)^{2n} \int \{f(x)\}^2 \, dx$$

This latter formula applies to the principal solution.

(b) Approximation of Integrals

As indicated previously, it is possible to approximate to integral equations by sets of simultaneous equations. The basis for this is that it is possible to approximate to

$$\int_a^b P(y) \, dy$$

by an expression of the form

$$\sum_{\beta=1}^{n} c_\beta P(x_\beta)$$

where if

$$a = y_0 \leqslant y_1 < \ldots \qquad y_\beta < y_{\beta+1} \ldots \qquad \leqslant y_n = b$$

$$c_\beta = y_\beta - y_{\beta-1}, \qquad y_{\beta-1} \leqslant x_\beta \leqslant y_\beta$$

Examples of these are the formulae

$$\int_0^{2h} P(y) \, dy = \frac{h}{3}[P(0)+4P(h)+P(2h)]$$

or

$$\int_{-1}^{+1} P(y) \, dy = P(-1/\sqrt{3}) + P(1\sqrt{3})$$

Thus the integral equation

$$\phi(x) = \lambda \int K(x, y)\phi(y) \, dy + f(x)$$

is replaced by a set of equations

$$\phi(x_\alpha) = \lambda \sum_{\beta=1}^{n} c_\beta K(x_\alpha, x_\beta)\phi(x_\beta) + f(x_\alpha) \qquad 1 \leqslant \alpha \leqslant n$$

and the equation

$$\int K(x, y)\phi(y) \, dy = f(x)$$

is replaced by the set of equations

$$\sum_{\beta=1}^{n} c_\beta K(x_\alpha, x_\beta)\phi(x_\beta) = f(x_\alpha) \qquad 1 \leqslant \alpha \leqslant n \tag{5.15}$$

and the integral equations have been reduced to the solution of a set of n linear equations in n variables. It may be remarked that this method could also be applied to non-linear integral equations, but the solution of the equations may give rise to difficulties.

Example 5.5

Find approximations to $\phi(\frac{1}{4})$ and $\phi(\frac{3}{4})$ when $\phi(x)$ is determined by the integral equation

$$\phi(x) - \int_0^1 e^{xy}\phi(y)\,dy = 1 - x^{-1}(e^x - 1)$$

The integral

$$\int_0^1 g(x)\,dx$$

will be approximated by $\frac{1}{2}[g(\frac{1}{4}) + g(\frac{3}{4})]$. This is a crude approximation, but the idea of the method will be best seen by a very simple approximation. Thus

$$\int_0^1 e^{xy}\phi(y)\,dy$$

is replaced by

$$\tfrac{1}{2}[e^{x/4}\phi(\tfrac{1}{4}) + e^{3x/4}\phi(\tfrac{3}{4})]$$

and the integral equation is replaced by two linear equations for $\phi(\frac{1}{4})$ and $\phi(\frac{3}{4})$

$$\phi(\tfrac{1}{4}) - \tfrac{1}{2}[e^{1/16}\phi(\tfrac{1}{4}) + e^{3/16}\phi(\tfrac{3}{4})] = 1 - 4(e^{\frac{1}{4}} - 1)$$

$$\phi(\tfrac{3}{4}) - \tfrac{1}{2}[e^{3/16}\phi(\tfrac{1}{4}) + e^{9/16}\phi(\tfrac{3}{4})] = 1 - \tfrac{4}{3}(3^{\frac{3}{4}} - 1)$$

These equations become

$$\cdot 4768\,\phi(\tfrac{1}{4}) - \cdot 6031\,\phi(\tfrac{3}{4}) = -\cdot 1360$$

$$-\cdot 6031\,\phi(\tfrac{1}{4}) + \cdot 1245\,\phi(\tfrac{3}{4}) = -\cdot 4893$$

$$\phi(\tfrac{1}{4}) = 1\cdot 021, \qquad \phi(\tfrac{3}{4}) = 1\cdot 014$$

(The exact solution is $\phi(x) = 1$.)

(c) Approximation of Kernel and Free Term

In many cases, an approximation solution to an integral equation of the second kind can be determined by approximation to the kernel and to the free term, and conveniently expressions are available which give the error involved.

An approximate solution $\phi^*(x)$ to the integral equation

$$\phi(x) = \lambda \int K(x, y)\phi(y)\,\mathrm{d}y + f(x) \tag{5.16a}$$

is given by the solution of the integral equation

$$\phi^*(x) = \lambda \int K^*(x, y)\phi^*(y)\,\mathrm{d}y + f^*(x) \tag{5.16b}$$

where $K^*(x, y)$ and $f^*(x)$ are respectively approximations to $K(x, y)$ and $f(x)$, which will be taken so that they are easier to handle than the original $K(x, y)$ and $f(x)$. For example $K^*(x, y)$ might be degenerate and $f^*(x)$ might be the first few terms of a power series. The actual nature is not important. The error is obtained as follows. Let

$$\int |K(x, y) - K^*(x, y)|\,\mathrm{d}y < \varepsilon$$

$$|f(x) - f^*(x)| < \eta \quad |f(x) < M$$

and suppose also that the resolvent kernel $R^*(x, y, \lambda)$ associated with Eq. (5.16b) obeys the relation

$$\int |R^*(x, y, \lambda)|\,\mathrm{d}y < C$$

Let

$$r(x) = \phi^*(x) - \phi(x)$$

In order to obtain a bound for $r(x)$ it is first necessary to obtain a bound for $\phi(x)$. This is done as follows:

$$\phi(x) = \lambda \int K^*(x, y)\phi(y)\,\mathrm{d}y + g(x) \tag{5.17a}$$

where

$$g(x) = f(x) + \lambda \int \{K(x, y) - K^*(x, y)\}\phi(y)\,\mathrm{d}y \tag{5.17b}$$

whence

$$\phi(x) = g(x) + \lambda \int R^*(x, y, \lambda) g(y) \, dy \tag{5.17c}$$

Suppose that $\max|\phi(x)| = N$. It follows from Eq. (5.17b) that

$$|g(x)| \leqslant |f(x)| + |\lambda| \int |K(x, y) - K^*(x, y)| \, |\phi(y)| \, dy \leqslant M + |\lambda| \varepsilon N$$

It follows from Eq. (5.17c) that

$$|\phi(x)| \leqslant |g(x)| + |\lambda| \int |R^*(x, y, \lambda)| \, |g(y)| \, dy$$

Thus

$$N \leqslant (1 + |\lambda| C)(M + |\lambda| \varepsilon N)$$

and so

$$N \leqslant \frac{M(1 + |\lambda| C)}{1 - |\lambda| \varepsilon (1 + |\lambda| C)} \tag{5.18}$$

The division is in order if

$$1 - |\lambda| \varepsilon (1 + |\lambda| C) > 0$$

The result (5.18) implies that all solutions of the original integral equation are bounded, that the solution is unique, and that λ is not an eigenvalue of the original equation.

Now that a bound for $\phi(x)$ has been determined it is possible to estimate a bound for $r(x)$. It follows from Eqs. (5.16) and (5.17) that

$$r(x) - \lambda \int K^*(x, y) r(y) \, dy = g(x) - f(x) + f(x) - f^*(x) = h(x) \text{ say}$$

whence

$$r(x) = h(x) + \lambda \int R^*(x, y, \lambda) h(y) \, dy$$

Now

$$|h(x)| \leqslant |g(x) - f(x)| + |f(x) - f^*(x)| \leqslant |\lambda| \varepsilon N + \eta$$

and

$$|r(x)| \leqslant (1 + |\lambda| C)(|\lambda| \varepsilon N + \eta)$$

$$\leqslant (1 + |\lambda| C) \left\{ \frac{M |\lambda| \varepsilon (1 + |\lambda| C)}{1 - |\lambda| \varepsilon (1 + |\lambda| C)} + \eta \right\} \tag{5.19}$$

Alternatively if

$$|\phi^*(x)| \leqslant N^*, \qquad \text{then} \quad N \leqslant N^* + \rho$$

where

$$\max|r(x)| = \rho$$

Thus

$$\rho \leqslant (1+|\lambda|C)\{|\lambda|\varepsilon(N^*+\rho)+\eta\}$$

and so

$$\rho \leqslant \frac{(|\lambda|\varepsilon N^*+\eta)(1+|\lambda|C)}{1-|\lambda|\varepsilon(1+|\lambda|C)} \tag{5.20}$$

The caveat must of course be entered that the denominator is positive.

Example 5.6

Find an approximate solution to the integral equation

$$\phi(x) = \int_0^{\frac{1}{2}} \cos xy\phi(y)\,dy + f(x)$$

by replacing $\cos xy$ by

$$1 - \frac{x^2y^2}{2}$$

$\phi^*(x)$ is the solution of

$$\phi^*(x) = \int_0^{\frac{1}{2}} \left(1 - \frac{x^2y^2}{2}\right)\phi^*(y)\,dy + f(x)$$

and so

$$\phi^*(x) = A + Bx^2 + f(x)$$

If

$$\int_0^{\frac{1}{2}} f(y)\,dy = f_0 \qquad \int_0^{\frac{1}{2}} y^2 f(y)\,dy = f_2$$

it can be shown by the usual methods that

$$A = \frac{2889 f_0 - 60 f_2}{1497}$$

and

$$B = \frac{-60 f_0 - 720 f_2}{1497}$$

The magnitude of $r^*(x)$ may be estimated from the following data: $\lambda = 1$

$$\int_0^{\frac{1}{2}} \left| \cos xy - \left(1 - \frac{x^2 y^2}{2}\right) \right| dy \leqslant \int_0^{\frac{1}{2}} \frac{x^4 y^4}{4!} dy = \frac{x^4}{120} \leqslant \frac{1}{120} \cdot \frac{1}{16}$$

Thus

$$\varepsilon = \frac{1}{1920}$$

$$\phi^*(x) = \frac{1}{1497} \int_0^{\frac{1}{2}} (2889 - 60x^2 - 60y^2 - 720x^2 y^2) f(y)\, dy + f(x)$$

Thus

$$R(x, y, 1) = \frac{1}{1497} [2889 - 60x^2 - 60y^2 - 720x^2 y^2]$$

$$\int_0^{\frac{1}{2}} |R(x, y, 1)|\, dy = \frac{1}{1497} \left[\frac{2889 - 60x^2}{2} - \frac{(20 + 240x^2)}{8} \right]$$
$$< \tfrac{1}{2} \quad \text{if} \quad 0 \leqslant x \leqslant \tfrac{1}{2}$$

and so

$$C = \tfrac{1}{2}$$

Now

$$1 - |\lambda| \varepsilon (1 + |\lambda| C) = 1 - \frac{1}{1920}(1 + \tfrac{1}{2}) \simeq 1$$

and so

$$|r(x)| < \frac{M \dfrac{1}{1920}(1 + \tfrac{1}{2})}{1 - \dfrac{1}{1920}(1 + \tfrac{3}{2})} \simeq \frac{1}{1280} M$$

(d) Collocation

A method which can be used to find approximate solutions for Fredholm integral equations, both of the first kind and of the inhomogeneous second kind, is the method of collocation. This will be discussed here in terms of equations of the first kind, but the method is equally applicable to equations of the second kind. It may be remarked that the method can also be applied to non-linear integral equations, but this will not be discussed here.

Consider the function

$$R(\psi, x) = \int K(x, y)\psi(y)\,\mathrm{d}y - f(x) \tag{5.21}$$

If $\phi(x)$ is the solution of the integral equation

$$\int K(x, y)\phi(y)\,\mathrm{d}y = f(x)$$

then

$$R(\phi, x) = 0 \tag{5.22}$$

The idea behind the collocation method is that, if the function $R(\psi, x)$ can be made to vanish at a number of points, then $\psi(x)$ is an approximation to $\phi(x)$, and the more points at which $R(\psi, x)$ vanishes the better the approximation of $\psi(x)$ to $\phi(x)$. Suppose that $\psi_r(x)$, $1 \leqslant r \leqslant m$, form a linearly independent set of functions such as, for example, $x^r, 0 \leqslant r \leqslant m-1$. Let

$$\psi(x) = \sum_{r=1}^{m} C_r \psi_r(x)$$

Then

$$R(\psi, x) = \sum_{r=1}^{m} C_r \int K(x, y)\psi_r(y)\,\mathrm{d}y - f(x) \tag{5.23}$$

Suppose now that $R(\psi, x)$ vanishes at m points x_s which are within the domain of interest. Then

$$\sum_{r=1}^{m} C_r \int K(x_s, y)\psi_r(y)\,\mathrm{d}y - f(x_s) = 0 \qquad 1 \leqslant s \leqslant m \tag{5.24}$$

There are now m equations in m variables and so the C_r can be calculated, and an approximate solution can be found.

Example 5.7

Calculate an approximation of the form $a_0 + a_1 y$ to $\phi(y)$ when $\phi(y)$ is given by the integral equation

$$\int_0^1 \mathrm{e}^{xy}\phi(y)\,\mathrm{d}y = (x+1)^{-1}[\mathrm{e}^{x+1} - 1] \qquad 0 \leqslant x \leqslant 1$$

Let

$$R(\psi, x) = \int_0^1 \mathrm{e}^{xy}\psi(y)\,\mathrm{d}y - (x+1)^{-1}[\mathrm{e}^{x+1} - 1]$$

Then

$$R(\phi, x) = 0$$

Suppose that the coefficients a_0 and a_1 are determined by the equations

$$R(\psi, 0) = 0$$

$$R(\psi, 1) = 0$$

$$R(\psi, 0) = \int_0^1 (a_0 + a, y)\,dy - (e-1)$$

$$= a_0 + \tfrac{1}{2}a_1 - (e-1)$$

$$R(\psi, 1) = \int_0^1 e^y(a_0 + a_1 y)\,dy - \tfrac{1}{2}(e^2 - 1)$$

$$= a_0(e-1) + a_1 - \tfrac{1}{2}(e^2 - 1)$$

Solving

$$a_0 = \frac{e-1}{2}, \qquad a_1 = e - 1$$

and

$$\psi(x) = \tfrac{1}{2}(e-1)(1+2x)$$

This is to be compared with an exact solution $\phi(x) = e^x$.

(e) Galerkin's Method

The idea behind Galerkin's approach to an approximate solution of the integral equation

$$R(\phi, x) \equiv \int K(x, y)\phi(y)\,dy - f(x) = 0$$

is that

$$\int R(\phi, x)\chi(x)\,dx = 0$$

for any function $\chi(x)$ whatsoever. Thus if

$$\int R(\psi, x)\chi_s(x)\,dx = 0 \tag{5.25}$$

for a number of linearly independent functions $\chi_s(x)$, $\psi(x)$ is an approximation in some sense to $\phi(x)$ and the more functions to which $R(\psi, x)$ is orthogonal, the better the approximation $\psi(x)$ is to $\phi(x)$. If there are m such linearly independent functions χ_s, it will be desirable to have m constants to be determined in ψ. Suppose that $\psi_r(x), 1 \leqslant r \leqslant m$ form a linearly independent set of functions and assume that

$$\psi(x) = \sum_{r=1}^{m} C_r \psi_r(x)$$

The sets of functions $\psi_r(x)$ and $\chi_s(x)$ can always, from the point of view of the theory, be assumed to be normalized and orthogonalized. Equation (5.25) can be written as

$$\sum_{r=1}^{m} \int\int C_r \psi_r(y) K(x, y) \chi_s(x) \,\mathrm{d}x \,\mathrm{d}y - \int f(x) \chi_s(x) \,\mathrm{d}x = 0 \tag{5.26}$$

that is

$$\sum_{r=1}^{m} \gamma_{rs} C_r = f_s, \qquad 1 \leqslant s \leqslant m \tag{5.27}$$

where

$$\gamma_{rs} = \int\int \psi_r(y) K(x, y) \chi_s(x) \,\mathrm{d}x \,\mathrm{d}y$$

$$f_s = \int f(x) \chi_s(x) \,\mathrm{d}x$$

The set of equations (5.27) can be solved for the C_r, and an approximate solution for $\phi(x)$ follows.

The case of an integral equation of the second kind can also be discussed in this way. In this case, the relation to be satisfied is

$$\int\left[\psi(x) - \lambda \int K(x, y) \psi(y) \,\mathrm{d}y - f(x)\right] \chi_s(x) \,\mathrm{d}x = 0, \qquad 1 \leqslant s \leqslant m$$

Let

$$\psi(x) = f(x) + \sum_{r=1}^{m} C_r \psi_r(x) \tag{5.28}$$

Let

$$\int \psi_r(x) \chi_s(x) \,\mathrm{d}x = \beta_{rs}$$

and let

$$\iint \chi_s(x) K(x, y) f(y) \,\mathrm{d}x \,\mathrm{d}y = k_s$$

Then the relation (5.27) becomes

$$\sum_{r=1}^{m} (\beta_{rs} - \lambda \gamma_{rs}) C_r = \lambda k_s, \qquad 1 \leqslant s \leqslant m \tag{5.29}$$

It may be remarked that this process is equivalent to replacing $K(x, y)$ by a degenerate kernel. Let

$$V_s(y) = \int K(x, y)\chi_s(x)\,dx$$

and consider the integral equation

$$\psi(x) - \lambda \int K^{(m)}(x, y)\psi(y)\,dy = f(x) \tag{5.30}$$

where

$$K^{(m)}(x, y) = \sum_{i=1}^{m} \chi_i(x)V_i(y) \tag{5.31}$$

It can be seen that

$$\begin{aligned}\int\int K^{(m)}(x, y)\psi(y)\chi_s(x)\,dx\,dy &= \int\int \sum_{i=1}^{m} \chi_i(x)V_i(y)\psi(y)\chi_s(x)\,dx\,dy \\ &= \int V_s(y)\psi(y)\,dy \\ &= \int\int K(x, y)\psi(y)\chi_s(x)\,dx\,dy\end{aligned}$$

Thus the solution $\psi(x)$ as defined by Eqs. (5.28) and (5.30) is the solution of the degenerate kernel integral equation (5.31), and the results of Section 5.4(c) associated with bounds may be used.

Example 5.8

Find a two term approximation to the solution of the integral equation

$$\phi(x) - \int_0^1 K(x, y)\phi(y)\,dy = x, \qquad 0 \leqslant x \leqslant 1$$

where

$$\begin{aligned}K(x, y) &= x, \qquad x \leqslant y \\ &= y, \qquad x \geqslant y\end{aligned}$$

using Galerkin's method.

The equation can be rewritten in the form

$$\phi(x) - \int_0^x y\phi(y)\,dy - x\int_x^1 \phi(y)\,dy - x = 0$$

Look for an approximate solution of the form $\psi(x) = a_0 + a_1 x$. {In this case $f(x) = x$ and so need not appear explicitly in $\psi(x)$.}

$$\int_0^x y\psi(y)\,\mathrm{d}y = a_0\frac{x^2}{2} + \frac{a_1 x^3}{3}$$

$$\int_x^1 \psi(y)\,\mathrm{d}y = a_0(1-x) + \frac{a_1}{2}(1-x^2)$$

Thus

$$\psi(x) - \int_0^1 K(x,y)\psi(y)\,\mathrm{d}y = a_0\left(1 - x + \frac{x^2}{2}\right) + a_1\left(\frac{x}{2} + \frac{1}{6}x^3\right)$$

It follows, taking $\chi_1(x) = 1$, and $\chi_2(x) = x$ that

$$\int_0^1 [R(\psi,x) - f(x)]\,\mathrm{d}x = \tfrac{2}{3}a_0 + \tfrac{7}{24}a_1 - \tfrac{1}{2} = 0$$

and

$$\int_0^1 [R(\psi,x) - f(x)]x\,\mathrm{d}x = \tfrac{7}{24}a_0 + \tfrac{1}{5}a_1 - \tfrac{1}{3} = 0$$

whence

$$a_0 = \tfrac{8}{139} = 0{\cdot}0576$$

$$a_1 = \tfrac{220}{139} = 1{\cdot}5827$$

The exact solution is sec 1 sin x.

(f) Least Squares Method

Suppose that $\phi(x)$ is the function which is the solution of the integral equation

$$R(\psi,x) \equiv \int K(x,y)\psi(y)\,\mathrm{d}y - f(x) = 0$$

Then if $\omega(x)$ is positive everywhere in the domain of integration

$$I(\psi) = \int [R(\psi,x)]^2\omega(x)\,\mathrm{d}x$$

is positive for all functions $\psi(x)$ except $\phi(x)$, in which case it is zero. Thus the quality of the approximation ψ to ϕ may be measured by the smallness of I and a diminution of $I(\psi)$ is equivalent to improving the approximation.

Consider an approximation of the form

$$\psi(x) = \sum_{r=1}^{n} c_r \psi_r(x) \tag{5.32}$$

where the ψ_r are linearly independent and may be assumed to be normalized and orthogonalized. It is then possible to write

$$I(\psi) = J(c_1, \ldots, c_n)$$

and it is necessary to consider how J may be made as small as possible. Finding the values of the c_r which minimize J is equivalent to finding the best approximation to $\phi(x)$ of the form of Eq. (5.32) with weighting function $\omega(x)$.

$$R(\psi, x) = \sum_{r=1}^{n} c_r \int K(x, y)\psi_r(y)\,\mathrm{d}y - f(x)$$

$$J(c_1, \ldots, c_n) = \int \left[\sum_{r=1}^{n} c_r \int K(x, y)\psi_r(y)\,\mathrm{d}y - f(x)\right]^2 \omega(x)\,\mathrm{d}x$$

The value of J which is a minimum is given by the c_r being the solutions of the n equations

$$\frac{\partial J}{\partial c_s} = 0, \qquad 1 \leqslant s \leqslant n$$

that is

$$\int \left\{\sum_{r=1}^{n} c_r \int K(x, y)\psi_r(y)\,\mathrm{d}y - f(x)\right\} \left\{\int K(x, z)\psi_s(z)\,\mathrm{d}z\right\} \omega(x)\,\mathrm{d}x = 0, \qquad 1 \leqslant s \leqslant n$$

Let

$$\left[\int K(x, z)\psi_s(z)\,\mathrm{d}z\right]\omega(x) = \chi_s(x)$$

Then the c_r are defined by

$$\int \left[\sum_{r=1}^{n} c_r \int K(x, y)\psi_r(y)\,\mathrm{d}y - f(x)\right] \chi_s(x)\,\mathrm{d}x = 0 \tag{5.33}$$

that is

$$\int R(\psi, x)\chi_s(x)\,\mathrm{d}x = 0, \qquad 1 \leqslant s \leqslant n$$

Thus the least squares method for finding an approximate solution leads to equations which are identical in form with those obtained using Galerkin's method. Consequently it is not worth working an example. It is however worth noting that it is sometimes possible to use the least squares method for a non-linear integral equation.

Example 5.9

Find an approximate solution of the form $a+bx^2$ to the integral equation

$$\cos \pi x \int_0^1 \{\phi(y)\}^2 \mathrm{d}y = \phi(x), \qquad 0 \leqslant x \leqslant 1 \tag{5.34}$$

Let

$$J(\psi) = \int_0^1 \left[\cos \pi x \int_0^1 \left[\psi(y) \right]^2 \mathrm{d}y - \psi(x) \right]^2 \mathrm{d}x$$

$J(\psi) > 0$ unless $\psi = \phi$ when $J(\phi) = 0$

$$J(\psi) = \tfrac{1}{2}\left[\int_0^1 [\psi(y)]^2 \mathrm{d}y \right]^2 - 2\int_0^1 \cos \pi x \psi(x) \int_0^1 [\psi(y)]^2 \mathrm{d}y \mathrm{d}x + \int_0^1 [\psi(x)]^2 \mathrm{d}x$$

$$= \int_0^1 [\psi(y)]^2 \mathrm{d}y \left\{ \tfrac{1}{2} \int_0^1 [\psi(y)]^2 \mathrm{d}y - 2\int_0^1 \psi(x) \cos \pi x \, \mathrm{d}x + 1 \right\}$$

$$= \int_0^1 [\psi(y)]^2 \mathrm{d}y I(\psi)$$

say where $I(\psi) > 0$ unless $\psi = \phi$ when $I(\phi) = 0$. If $\psi(x) = a+bx^2$, it can be shown that

$$I(\psi) = \frac{1}{2}\left(a^2 + \frac{2ab}{3} + \frac{b^2}{5} \right) + \frac{4b}{2\pi^2} + 1 > 0$$

This is a minimum with respect to a and b if

$$a + \frac{b}{3} = 0, \qquad \frac{a}{3} + \frac{b}{5} + \frac{4}{\pi^2} = 0$$

whence

$$\psi(x) = \frac{15}{\pi^2}(1 - 3x^2)$$

which may be compared with the actual solution $2\cos \pi x$.

(g) Schwinger's Principle

Suppose that a quantity Q_0 is defined by

$$Q_0 = \int f(x)\phi(x) \mathrm{d}x$$

where $f(x)$ is a known function and $\phi(x)$ is the solution of the integral equation of the first kind.

$$f(x) = \int K(x, y)\phi(y) \mathrm{d}y$$

$K(x, y)$ is positive definite, and an approximate value for Q_0 is needed. Alternative expressions for Q_0 are

$$\int\int \phi(x)K(x, y)\phi(y)\,\mathrm{d}y$$

and

$$2\int f(x)\phi(x)\,\mathrm{d}x - \int \phi(x)K(x, y)\phi(y)\,\mathrm{d}y$$

Define a functional

$$Q^*(\alpha\psi) = 2\alpha\int f(x)\psi(x)\,\mathrm{d}x - \alpha^2\int\int \psi(x)K(x, y)\psi(y)\,\mathrm{d}y$$

α is a constant which is unspecified

$$Q^*(\phi) = Q_0$$

Let

$$\alpha\psi(x) = \phi(x) + \chi(x)$$

$$Q^*(\alpha\psi) = 2\int f(x)\{\phi(x)+\chi(x)\}\,\mathrm{d}x - \int\int \phi(x)K(x, y)\phi(y)\,\mathrm{d}x\,\mathrm{d}y$$

$$-2\int\int \chi(x)K(x, y)\phi(y)\,\mathrm{d}x\,\mathrm{d}y - \int\int \chi(x)K(x, y)\chi(y)\,\mathrm{d}x\,\mathrm{d}y$$

$$= Q_0 - \int\int \chi(x)K(x, y)\chi(y)\,\mathrm{d}x\,\mathrm{d}y < Q_0 \qquad (5.35)$$

This follows because K is symmetric and positive definite and because $\phi(x)$ satisfies the integral equation.

Thus $Q^*(\alpha\psi)$ is an approximate expression for Q_0 and the error is of the second order in $\chi = \phi - \alpha\psi$.

Now $Q^*(\alpha\psi) < Q_0$ always, and so the best approximation will be the one which maximizes $Q^*(\alpha\psi)$. α was an unspecified constant, and $Q^*(\alpha\psi)$ is a quadratic expression in α which is maximized when α is given by the relation

$$\frac{\partial Q^*}{\partial \alpha} = 0$$

that is

$$\alpha = \alpha_{\max} = \frac{\int f(x)\psi(x)\,\mathrm{d}x}{\int\int \psi(x)K(x, y)\psi(y)\,\mathrm{d}x\,\mathrm{d}y}$$

Thus the quantity

$$Q(\psi) = Q^*(\alpha_{\max}\psi)$$

$$= \frac{\left[\int f(x)\psi(x)\,\mathrm{d}x\right]^2}{\int\int \psi(x)K(x,y)\psi(y)\,\mathrm{d}x\,\mathrm{d}y}$$

is always less than Q_0 (unless $\psi = \phi$) and will differ from Q_0 by a small quantity of the second order. It will be seen that $Q(\psi)$ is independent of the scale of ψ. It may be convenient, if better approximations are needed, to introduce parameters into ψ and maximize $Q(\psi)$ with respect to them.

Example 5.10

Find a lower bound to the integral

$$\int_{-\pi/2}^{\pi/2} \left(2+\frac{\pi}{2}\cos x\right)\phi(x)\,\mathrm{d}x \tag{5.36}$$

when $\phi(x)$ is defined by the integral equation

$$2+\frac{\pi}{2}\cos x = \int_{-\pi/2}^{\pi/2} (1+\cos x\cos y)\phi(y)\,\mathrm{d}y$$

Let

$$Q(\psi) = \frac{\left[\int_{-\pi/2}^{\pi/2} \left(2+\frac{\pi}{2}\cos x\right)\psi(x)\,\mathrm{d}x\right]^2}{\int_{-\pi/2}^{\pi/2}\int_{-\pi/2}^{\pi/2} \psi(x)(1+\cos x\cos y)\psi(y)\,\mathrm{d}x\,\mathrm{d}y}$$

Then $Q(\psi)$ provides a lower bound to the integral. If $\psi(x) = 1$, it can be shown that

$$Q(\psi) = (3\pi)^2/(\pi^2+4) = 6{\cdot}11$$

$\phi(x) = \cos x$ and the true value is

$$4+\frac{\pi^2}{4} = 6{\cdot}47$$

5.5 Approximate Evaluation of Eigenvalues and Eigenfunctions

(a) Some Preliminary Remarks

This treatment will be confined to cases where the kernel is of the form

$$K(x,y) = \sum_{s=1}^{\infty} \frac{\phi_s(x)\phi_s(y)}{\lambda_s} \tag{5.37}$$

where the ϕ_s form a complete orthonormal set and the λ_s are arranged so that

$$|\lambda_n| < |\lambda_{n+1}| \qquad \text{for all } n$$

The following formulae for the trace of the nth iterated kernel will be useful in considering the approximation of eigenvalues and eigenfunctions

$$\mathrm{Tr} K_n = \sum_{s=1}^{\infty} \lambda_s^{-n} \tag{5.38}$$

If

$$f(x) = \sum_{s=1}^{\infty} f_s \phi_s(x) \, \mathrm{d}x \qquad \text{where} \qquad f_s = \int f(y)\phi_s(y) \, \mathrm{d}y \tag{5.39}$$

then

$$f_n(x) = \int K_n(x, z) f(z) \, \mathrm{d}z = \sum_{s=1}^{\infty} \frac{f_s}{\lambda_s^n} \phi_s(x) \tag{5.40}$$

where K_n is the nth iterated kernel.

It will be assumed almost everywhere that it is only the first eigenvalue λ_1 and the first eigenfunction $\phi_1(x)$ which are of interest. If higher eigenvalues and eigenfunctions are of interest this will be stated explicitly. It will also be assumed that the eigenvalues are well separated. The implication of this will be that the quantities

$$\left(\frac{\lambda_n}{\lambda_{n+1}}\right)^s \qquad \text{and still more} \qquad \left(\frac{\lambda_n}{\lambda_{n+p}}\right)^s, \qquad p > 1$$

are negligible for some reasonably small number, s.

(b) Kellog's Iterative Process

Suppose that $f(x)$ is some arbitrary function. Consider the sequence of functions

$$f_n(x) = \int K(x, y) f_{n-1}(y) \, \mathrm{d}y, \qquad n > 0$$

$$f_0(x) = f(x)$$

From Eq. (5.40)

$$f_n(x) = \sum_{s=1}^{\infty} f_s \lambda_s^{-n} \phi_s(x)$$

If n is sufficiently large, it is possible to write

$$f_n(x) \sim f_1 \lambda_1^{-n} \phi_1(x) \tag{5.41a}$$

and

$$f_{n+1}(x) \sim f_1 \lambda_1^{-(n+1)} \phi_1(x) \tag{5.41b}$$

Thus an approximate estimate for λ_1 is given by $f_n(x)/f_{n+1}(x)$ and

$$\lambda_1 = \lim_{n\to\infty} f_n(x)/f_{n+1}(x)$$

Clearly $f_{n+1}(x)$ will not be an exact constant multiple of $f_n(x)$ but when n is sufficiently large the quantity $f_n(x)/f_{n+1}(x)$ will not differ too much from a constant and so an estimate can be made. A possible estimate is

$$\left\{ \frac{\int [f_n(x)]^2 \,\mathrm{d}x}{\int [f_{n+1}(x)]^2 \,\mathrm{d}x} \right\}^{\frac{1}{2}} \tag{5.42}$$

The sign being obvious in the context.

An approximation to the first eigenfunction can be determined by the relation

$$\phi_1(x) \sim \frac{f_n(x)}{\int [f_n(x)]^2 \mathrm{d}x} \tag{5.43}$$

The advantage of this method is that numerical integration can be used when K is defined only by a set of data.

The second eigenvalue and eigenfunction can be obtained as follows. The quantity

$$f^*(x) = f(x) - \left[\int f(y)\phi_1(y)\,\mathrm{d}y \right] \phi_1(x) \tag{5.44}$$

$$= \sum_{s=2}^{\infty} f_s \phi_s(x)$$

does not so to speak contain any element of $\phi_1(x)$, and a procedure similar to the previous one may be used to obtain λ_2 and $\phi_2(x)$. The process may be followed through as often as required. It should be noted that as in numerical integration small errors can occur, it is possible for a small amount of $\phi_1(x)$ to get back into the calculations so to speak, and eventually swamp $\phi_2(x)$. Thus it is advisable to 'purify' $f_n^*(x)$ by replacing it by

$$f_n^*(x) - \left[\int f_n^*(y)\phi_1(y)\,\mathrm{d}y \right] \phi_1(x)$$

It may be remarked that if accidentally $f(x) = K\phi_s(x)$ it will be ϕ_s and λ_s which emerge by this process. Thus care should be taken if possible for the initial $f(x)$ to have a qualitative behaviour similar to that which might be expected of $\phi_1(x)$. It can be seen also that the following iterative processes will converge to λ_1:

$$g_n = \frac{\int [f_n(x)]^2\,\mathrm{d}x}{\int\int f_n(x)K(x,y)f_n(y)\,\mathrm{d}x\,\mathrm{d}y} = \frac{\int [f_n(x)]^2\,\mathrm{d}x}{\int f_n(x)f_{n+1}(x)\,\mathrm{d}x}$$

$$= \left(\sum_{s=1}^{\infty} f_s^2/\lambda_s^{2n}\right)\bigg/\left(\sum_{s=1}^{\infty} f_s^2/\lambda_s^{2n+1}\right) \tag{5.45}$$

and

$$h_n = \frac{\int f_n(x)f_{n+1}(x)\,\mathrm{d}x}{\int [f_{n+1}(x)]^2\,\mathrm{d}x}$$

$$= \left(\sum_{s=1}^{\infty} f_s^2/\lambda_s^{2n+1}\right)\bigg/\left(\sum_{s=1}^{\infty} f_s^2/\lambda_s^{2n+2}\right) \tag{5.46}$$

Example 5.11

Find an approximation by Kellog's method to the value of the smallest eigenvalue of the kernel

$$\begin{aligned} K(x,y) &= x, \qquad 0 \leqslant x \leqslant y \leqslant 1 \\ &= y, \qquad 0 \leqslant y \leqslant x \leqslant 1 \end{aligned}$$

The integral equation is thus

$$\phi(x) = \lambda\left[\int_0^x y\phi(y)\,\mathrm{d}y + x\int_x^1 \phi(y)\,\mathrm{d}y\right]$$

It will be seen that $\phi(0) = 0$. Consequently a suitable function to start with is $f(x) = x$. It follows that the function sequence $f_n(x)$ has the following forms:

$$f_1(x) = \frac{1}{2}x - \frac{x^3}{6}$$

$$f_2(x) = \frac{5}{24}x - \frac{1}{12}x^3 + \frac{1}{120}x^5$$

$$f_3(x) = \frac{61}{720}x - \frac{5}{144}x^3 + \frac{1}{240}x^5 - \frac{1}{5040}x^7$$

Now

$$\lambda_1 = \lim_{n\to\infty} g_n(x) \qquad \text{where} \qquad g_n(x) = f_{n-1}(x)/f_n(x)$$

Consider the two sequences of approximations to λ_1 which are defined by

$$g_n(0) = \lim_{x\to 0} g_n(x) \qquad \text{and} \qquad g_n(1)$$

It can be seen that

$n = 1$	2	3
$g_n(0) = 2$	$2\frac{2}{5}$	$2\frac{28}{61} = 2{\cdot}426$
$g_n(1) = 3$	$2\frac{1}{2}$	$2\frac{24}{51} = 2{\cdot}471$

It can be verified that the actual eigenvalues are

$$\lambda_n = \left\{\frac{(2n-1)\pi}{2}\right\}^2$$

and the eigenfunctions are

$$\sin(2n-1)\frac{\pi x}{2}$$

The first eigenvalue is thus $\pi^2/4 = 2{\cdot}4674$ and the first eigenfunction $\sin \pi x/2$.

The reader is invited to evaluate the successive approximations

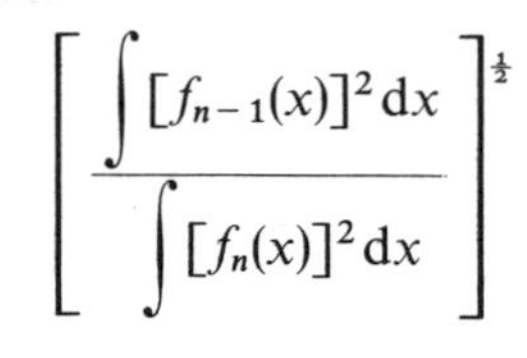

$$\left[\frac{\int [f_{n-1}(x)]^2\,dx}{\int [f_n(x)]^2\,dx}\right]^{\frac{1}{2}}$$

and compare

$$\frac{f_3(x)}{\left\{\int_0^1 [f_3(x)]^2\,dx\right\}^{\frac{1}{2}}} \qquad \text{with} \qquad \sin\frac{\pi x}{2}$$

(c) Use of Trace for Estimating First Eigenvalue

From Eq. (5.38), it follows that

$$\mathrm{Tr}K_{2n} = \sum_{s=1}^{\infty} \lambda_s^{-2n}$$

and

$$\mathrm{Tr}K_{2n-2} = \sum_{s=1}^{\infty} \lambda_s^{-(2n-2)}$$

Because $|\lambda_1| \leqslant |\lambda_s|$, $s \geqslant 2$, it follows that

$$\lambda_1^2 \leqslant \frac{\mathrm{Tr}K_{2n-2}}{\mathrm{Tr}K_{2n}} \tag{5.47}$$

As n becomes large the sign of $\mathrm{Tr}\,K_n$ will be dominated by that of $\lambda_1{}^{-n}$, and so the sign of λ_1 will be the same as that of $\mathrm{Tr}\,K_n/\mathrm{Tr}\,K_{n-1}$ for sufficiently large n. Also,

$$\mathrm{Tr}\,K_{2n} \leqslant \lambda_1{}^{-2n}$$

and so

$$\lambda_1^2 \geqslant (\mathrm{Tr}\,K_{2n})^{-1/n} \tag{5.48}$$

Between Eqs. (5.47) and (5.48) it is thus possible to find upper and lower bounds for λ_1^2. It may be remarked that if K is positive definite all the eigenvalues are positive and

$$\mathrm{Tr}\,K > \lambda_1^{-1}, \qquad \lambda_1 > (\mathrm{Tr}\,K)^{-1}$$

If λ_1 and λ_2 are reasonably well separated from λ_3, an approximation to λ_2 can be obtained as follows. If all eigenvalues above λ_2 be ignored

$$\mathrm{Tr}\,K_n = \frac{1}{\lambda_1^n} + \frac{1}{\lambda_2^n} \tag{5.49}$$

i.e.

$$\mathrm{Tr}\,K_n - \frac{1}{\lambda_1^n} = \frac{1}{\lambda_2^n}$$

also

$$\mathrm{Tr}\,K_m - \frac{1}{\lambda_1^m} = \frac{1}{\lambda_2^m}$$

and so

$$\frac{\mathrm{Tr}\,K_n - \dfrac{1}{\lambda_1^n}}{\mathrm{Tr}\,K_n - \dfrac{1}{\lambda_1^m}} = \lambda_2^{m-n} \tag{5.50}$$

λ_1 will have been determined and, if m and n are sufficiently large, Eq. (5.50) holds. If $m-n$ is odd, the sign of λ_2 also follows. This idea extends to the determination of higher eigenvalues.

Example 5.12

Find upper and lower bounds to the first eigenvalue of the kernel

$$\frac{1-\alpha^2}{1-2\alpha\cos(x-y)+\alpha^2} \qquad -\pi \leqslant x, y \leqslant \pi \qquad 0 < \alpha < 1$$

It may be shown by contour integration that

$$\int_{-\pi}^{\pi} \frac{(1-\alpha^2)(1-\beta^2)\,\mathrm{d}z}{\{1-2\alpha\cos(x-z)+\alpha^2\}\{1-2\beta\cos(y-z)+\beta^2\}}$$

$$= 2\pi\frac{1-\alpha^2\beta^2}{1-2\alpha\beta\cos(x-y)+\alpha^2\beta^2} \qquad 0<\alpha,\beta<1$$

Repeated application of this formula gives

$$K_n(x,y) = \frac{(2\pi)^{n-1}(1-\alpha^{2n})}{1-2\alpha^n\cos(x-y)+\alpha^{2n}}$$

$$\mathrm{Tr}\,K_n = \int_{-\pi}^{\pi} K_n(x,x)\,\mathrm{d}x = (2\pi)^n(1+\alpha^n)$$

For large n $\mathrm{Tr}\,K_n$ is always positive and so λ_1 is positive

$$\lambda_1^2 \geqslant \frac{1}{(\mathrm{Tr}\,K_{2n})}^{-1/n} = (2\pi)^{-2}(1+\alpha^{2n})^{-1/n}$$

also

$$\lambda_1^2 \leqslant \frac{\mathrm{Tr}\,K_{2n-2}}{\mathrm{Tr}\,K_{2n}} = \frac{1}{(2\pi)^2}\left(\frac{1+\alpha^{2n-2}}{1+\alpha^{2n}}\right)$$

In this case making n tend to infinity, it follows that

$$\lambda_1^2 = \frac{1}{(2\pi)^2} \quad \text{and so} \quad \lambda_1 = \frac{1}{2\pi}$$

It is not of course always obvious whether the limiting process can be carried through.

(d) *Variational Expressions for Eigenvalues*

Suppose that λ_n and $\phi_n(x)$ are the eigenvalues and eigenfunctions of the integral equation

$$\phi(x) = \lambda\int K(x,y)\phi(y)\,\mathrm{d}y$$

and that any function $\psi(x)$ can be expressed as

$$\psi(x) = \sum_{s=1}^{\infty} c_s\phi_s(x)$$

Then the functional $\Lambda(\psi)$ defined by

$$[\Lambda(\psi)]^{-1} = \frac{\int\int \psi(x)K(x, y)\psi(y)\,\mathrm{d}y}{\int [\psi(x)]^2\,\mathrm{d}x}$$

$$= \left(\sum_{s=1}^{\infty} c_s^2/\lambda_s\right) \Big/ \left(\sum_{s=1}^{\infty} c_s^2\right)$$

$$= \lambda_p^{-1} + \left(\sum_{\substack{s=1 \\ s\neq p}}^{\infty} c_s^2/(\lambda_s^{-1} - \lambda_p^{-1})\right) \Big/ \left(\sum_{s=1}^{\infty} c_s^2\right) \tag{5.51}$$

$\Lambda(\psi)$ is independent of the scale of ψ. Two consequences follow:

(a) If K is a positive definite kernel

$$[\Lambda(\psi)] > \lambda_1. \quad \text{as} \quad \lambda_s > \lambda_1, s > 1$$

(b) If $\psi(x)$ is almost $\phi_p(x)$, then the $c_s (s \neq p)$ will be small in comparison with c_p which may be taken as unity. Thus an approximation $\psi(x)$ to $\phi_p(x)$ will give a value for $\Lambda(\psi)$ which is near to λ_p and differs from it by a quantity of the second order.

With these ideas in mind, consider the approximation of $\phi(x)$ by a series of the form

$$\sum_{s=1}^{n} a_s \psi_s(x) \tag{5.52}$$

where the ψ_s are a set of linearly independent functions. It is convenient, but not necessary, to take them as the first n members of a complete orthonormal set which obeys the relation

$$\int \psi_r(x)\psi_s(x)\,\mathrm{d}x = \delta_{rs}$$

An alternative system is the set of functions defined by

$$f_r(x) = \int K_r(x, y) f(y)\,\mathrm{d}y \qquad n \geqslant r \geqslant 1$$

where $f(x)$ is arbitrary. Unless $f(x)$ is actually an eigenfunction, the f_r will be linearly independent, and they will furthermore obey the appropriate boundary conditions.

The evaluation of the eigenvalues and eigenfunction of the integral equation

$$\phi(x) = \lambda \int K(x, y)\phi(y)\,\mathrm{d}y \tag{5.53}$$

is equivalent to finding stationary values of the functional $\Lambda(\psi)$. Now if

$$\psi(x) = \sum_{s=1}^{n} a_s \psi_s(x)$$

$$\Lambda(\psi) \quad \frac{\int \sum_{r=1}^{n} a_r \psi_r(x) \sum_{s=1}^{n} a_s \psi_s(x)\,dx}{\iint \sum_{r=1}^{n} a_r \psi_r(x) K(x,y) \sum_{s=1}^{n} a_s \psi_s(y)\,dx\,dy}$$

$$= \frac{\sum_{r=1}^{n} \sum_{s=1}^{n} a_r a_s P_{rs}}{\sum_{r=1}^{n} \sum_{s=1}^{n} a_r a_s Q_{rs}} \tag{5.54}$$

where

$$P_{rs} = \int \psi_r(x)\psi_s(x)\,dx$$

$$Q_{rs} = \iint \psi_r(x) K(x,y) \psi_s(y)\,dx\,dy$$

The condition that $\Lambda(\psi)$ is stationary and has the value λ is given by

$$\sum_{r=1}^{n} (\lambda Q_{rs} - P_{rs})\, a_s = 0 \qquad 1 \leqslant r \leqslant n \tag{5.55}$$

The condition that there is a non-trivial solution for the a_s is that

$$|\lambda Q_{rs} - P_{rs}| = 0 \tag{5.56}$$

there are n roots to this equation $\lambda_t^{(n)}$ $1 \leqslant t \leqslant n$, and that there is an associated eigenvector $[a_s^{(t)}]$ which corresponds to a function

$$\Psi_t^{(n)}(x) = \sum_{s=1}^{n} a_s^{(t)} \psi_s(x) \tag{5.57}$$

It may be shown that if λ_t and ϕ_t are an eigenvalue and the corresponding eigenfunction of the original integral equation,

$$\lim_{n\to\infty} \lambda_t^{(n)} = \lambda_t \tag{5.58}$$

$$\lim_{n\to\infty} \Psi_t^{(n)} = \phi_t(x) \tag{5.59}$$

when $\Psi_t^{(n)}$ has been normalized. The proof of these statements is long and involves sophisticated analysis (see Reference 7).

It may be remarked that because of property (b) derived from Eq. (5.51) it may not be necessary to take many terms in the series

$$\sum_{s=1}^{n} a_s\psi_s(x)$$

An approximation to $\phi_1(x)$ will give an approximate eigenvalue which is in error by a quantity of the second order.

Example 5.13

Find an approximation to the smallest eigenvalue associated with the kernel

$$\begin{aligned} K(x, y) &= x \qquad 0 \leqslant x \leqslant y \leqslant 1 \\ &= y \qquad 0 \leqslant y \leqslant x \leqslant 1 \end{aligned}$$

Now

$$\int K(x, y)\psi(y)\,\mathrm{d}y = \int_0^x y\psi(y)\,\mathrm{d}y + x\int_x^1 \psi(y)\,\mathrm{d}y$$

Take

$$\psi_1(x) = 1, \qquad \psi_2(x) = x$$

Then

$$\begin{aligned} P_{11} &= 1, \qquad P_{12} = 0, \qquad P_{22} = \tfrac{1}{3} \\ Q_{11} &= \tfrac{1}{3}, \qquad Q_{12} = \tfrac{1}{12}, \qquad Q_{22} = \tfrac{1}{30} \end{aligned}$$

The eigenvalue equation is

$$\begin{vmatrix} \dfrac{\lambda}{3} - 1 \ , & \dfrac{\lambda}{12} \\ \dfrac{\lambda}{12} \ , & \dfrac{\lambda}{30} - \dfrac{1}{3} \end{vmatrix} = 0$$

whence

$$\lambda = \tfrac{4}{3}(13 \pm \sqrt{124})$$

i.e. 2·4857 or 32·1807.
It will be seen that the first root is a reasonable approximation to the first eigenvalue 2·4674. The second eigenvalue is in fact 22·207. In general an approximation with n terms will give reasonable values only for the first $n-1$ eigenvalues.

EXERCISES

1. Solve the integral equation

$$\phi(x) = \int_0^x \frac{1+y^2}{1+[\phi(y)]^2}\,dy$$

2. Find the first three functions in the sequence of functions arising in the solution of the equation

$$\phi(x) = 1 + \int_0^x \{[\phi(y)]^{\frac{1}{2}} + y\}\,dy$$

3. Find a first and second approximation in the iterative non-trivial solution of the integral equation

$$\int_0^1 (x+y)^2[\phi(y)]^2\,dy = \phi(x)$$

4. Find the conditions to be imposed on the real quantities λ, a, b for the non-trivial solutions of the integral equation

$$\phi(x) = \lambda \int_{-\pi}^{\pi} (a\cos x\cos y + b\cos 2x\cos 2y)(\phi(y) + [\phi(y)]^3)\,dy$$

to be real.

5. Use an iterative procedure to show that the integral equation

$$\phi(x) = \int_0^x \frac{y\phi(y)\,dy}{1+[\phi(y)]^2}$$

does not have a non-trivial solution.

6. Find an approximation solution valid for small x of the form $a_0 + a_1x + a_2x^2$ for the integral equation

$$\phi(x) = \sin x + \int_0^1 \cos xy\phi(y)\,dy$$

and estimate the error involved.

7. Use the method of Section 5.3b to find approximate values of $\phi(\frac{1}{4})$ and $\phi(\frac{3}{4})$ when $\phi(x)$ is the solution of the integral equation in the previous exercise.

8. Find the error in the solution of the integral equation

$$\phi(x) = \int_0^1 \sin xy\phi(y) + f(x)$$

when $\sin z$ is approximated to by $z - z^3/6$.

9. Using the collocation method find an approximation of the form $a+by^2$ to the solution of the integral equation

$$\int_0^1 e^{-x^2y^2}\phi(y)\,dy = 2(1-x^2) \qquad 0 \leqslant x \leqslant 1$$

10. Using Galerkin's method, find a two term approximation to the integral equation

$$\phi(x) - \int_0^1 K(x,y)\phi(y)\,dy = x \qquad 0 \leqslant x \leqslant 1$$

where

$$\begin{aligned} K(x,y) &= x(1-y) \qquad x \leqslant y \\ &= y(1-x) \qquad x \leqslant y \end{aligned}$$

11. Using the least squares method, find the best solution, with weighting function $\omega(x) = 1$, to the integral equation

$$\int_0^1 K(x,y)\phi(y)\,dy = x^2 \qquad 0 \leqslant x \leqslant 1$$

where

$$\begin{aligned} K(x,y) &= x \qquad x \leqslant y \\ &= y \qquad x \geqslant y \end{aligned}$$

12. Find an approximate solution of the form $ax+bx^3$ to the integral equation

$$\sin \pi x \int_0^1 [\phi(y)]^2\,dy = \phi(x) \qquad 0 \leqslant x \leqslant 1$$

and compare it with the actual solution.

13. Find a lower bound to the integral

$$\int_0^1 x\phi(x)\,dx$$

where $\phi(x)$ is defined by the integral equation

$$x = \int_0^1 K(x,y)\phi(y)\,dy$$

and

$$\begin{aligned} K(x,y) &= x(1-y) \qquad x \leqslant y \\ &= y(1-x) \qquad x \geqslant y \end{aligned}$$

14. Find upper and lower bounds to the first eigenvalue and an approximation to the second eigenvalue of the kernel

$$K(x, y) = x \qquad 0 \leqslant x \leqslant y \leqslant 1$$
$$= y \qquad 0 \leqslant y \leqslant x \leqslant 1$$

15. Find approximations to the first two eigenvalues of the kernel

$$K(x, y) = \tfrac{1}{2}x(2-y) \qquad 0 \leqslant x \leqslant y \leqslant 1$$
$$= \tfrac{1}{2}y(2-x) \qquad 0 \leqslant y \leqslant x \leqslant 1$$

Appendices

(A) Repeated Integral

Let $I_n f(x)$, when n is zero or a positive integer, be defined as follows:

$$\begin{aligned} I_n f(x) &= \int_{x_0}^{x} \frac{(x-\xi)^{n-1}}{(n-1)!} f(\xi)\,\mathrm{d}\xi \qquad && n > 1 \\ &= \int_{x_0}^{x} f(\xi)\,\mathrm{d}\xi && n = 1 \\ &= f(x) && n = 0 \end{aligned}$$

Then if $n > 1$

$$\begin{aligned} \frac{\mathrm{d}}{\mathrm{d}x}\{I_n f(x)\} &= \frac{(x-x)^{n-1}}{(n-1)!} f(x) + \int_{x_0}^{x} \frac{(x-\xi)^{n-2}}{(n-2)!} f(\xi)\,\mathrm{d}\xi \\ &= I_{n-1} f(x) \end{aligned}$$

Also

$$I_n f(x_0) = 0 \qquad n \geqslant 1$$

Thus

$$\begin{aligned} I_n f(x) &= \int_{x_0}^{x} \mathrm{d}x_1 \int_{x_0}^{x_1} \mathrm{d}x_2 \ldots \int_{x_0}^{x_{n-1}} f(\xi)\,\mathrm{d}\xi \\ &= \left[\int_{x_0}^{x}\right]^n f(\xi)\,\mathrm{d}\xi \end{aligned}$$

(B) The Bunyakovskii–Cauchy–Schwarz Inequalities

There is a class of inequalities which are known after one of three mathematicians according to the origin of the writer. Consider the quantity

$$\int_a^b \{\alpha\phi(x) + \beta\psi(x)\}^2\,\mathrm{d}x$$

where all quantities are real. This is positive (unless $\alpha\phi(x)+\beta\psi(x)$ is zero).

Thus

$$\alpha^2 \int_a^b \phi^2 \, \mathrm{d}x + 2\alpha\beta \int_a^b \phi\psi \, \mathrm{d}x + \beta^2 \int_a^b \psi^2 \, \mathrm{d}x \geqslant 0$$

However the condition that

$$A\alpha^2 + 2H\alpha b + B\beta^2 \geqslant 0$$

is

$$H^2 \leqslant AB$$

Thus

$$\left\{ \int_a^b \phi\psi \, \mathrm{d}x \right\}^2 \leqslant \int_a^b \phi^2 \, \mathrm{d}x \int_a^b \psi^2 \, \mathrm{d}x$$

A similar inequality holds for the series

$$\sum_n a_n, \quad \sum_n b_n$$

$$\left(\sum_n a_n b_n \right)^2 \leqslant \left(\sum_n a_n^2 \right) \left(\sum_n b_n^2 \right)$$

For complex quantities, the inequalities take the forms

$$\left| \int_a^b \phi\bar{\psi} \, \mathrm{d}x \right|^2 \leqslant \left(\int_a^b |\phi|^2 \, \mathrm{d}x \right) \left(\int_a^b |\psi|^2 \, \mathrm{d}x \right)$$

and

$$\left| \sum_n a_n \bar{b}_n \right|^2 \leqslant \left(\sum_n |a_n|^2 \right) \left(\sum_n |b_n|^2 \right)$$

respectively.

(C) A Continuity Theorem

Let $K(x, y)$ be bounded and continuous in x in $c \leqslant x \leqslant d$ and let

$$\int_a^b |\psi(y)|^2 \, \mathrm{d}x \text{ exist}$$

Let

$$u(x) = \int_a^b K(x, y)\psi(y) \, \mathrm{d}y \qquad c \leqslant x \leqslant d$$

Then

$$u(x+h)-u(x) = \int_a^b [K(x+h,y)-K(x,y)]\psi(y)\,\mathrm{d}y$$

Using the inequality of Appendix B, it follows that

$$|u(x+h)-u(x)|^2 \leqslant \int_a^b |K(x+h,y)-K(x,y)|^2\,\mathrm{d}y \int_a^b |\psi(y)|^2\,\mathrm{d}y$$

It then follows that $|u(x+h)-u(x)|$ will tend to zero with h, thereby showing that u is continuous.

(D) Convergence of Neumann Series

Let

$$\{A(x)\}^2 = \int |K(x,y)|^2\,\mathrm{d}y$$

$$\{B(y)\}^2 = \int |K(x,y)|^2\,\mathrm{d}x$$

and

$$\int \{A(x)\}^2\,\mathrm{d}x = \int \{B(y)\}^2\,\mathrm{d}y = N^2$$

that is

$$\iint |K(x,y)|^2\,\mathrm{d}x\,\mathrm{d}y \text{ is bounded}$$

Now

$$|K_2(x,y)|^2 = \left|\int K(x,z)K(z,y)\,\mathrm{d}z\right|^2 \leqslant \{A(x)\}^2\{B(y)\}^2 \text{ by Appendix B}$$

$$\begin{aligned}
|K_3(x,y)|^2 &= \left|\int K(x,z)K_2(z,y)\,\mathrm{d}z\right|^2 \\
&\leqslant \int |K(x,z)|^2\,\mathrm{d}x \int |K_2(z,y)|^2\,\mathrm{d}z \\
&\leqslant \{A(x)\}^2 \int \{A(z)\}^2\{B(y)\}^2\,\mathrm{d}z \\
&= \{A(x)\}^2\{B(y)\}^2 N^2
\end{aligned}$$

and in general

$$|K_{n+2}(x, y)|^2 \leqslant \{A(x)\}^2\{B(y)\}^2 N^n \qquad n \geqslant 0$$

The series

$$\sum_{n=2}^{\infty} \lambda^{n-1} K_n(x, y)$$

is dominated by the series

$$A(x)B(y) \sum_{n=1}^{\infty} |\lambda|^{n-1} N^{n-2}$$

which is a geometrical series with ratio $|\lambda|N$. Thus the series

$$\sum_{n=2}^{\infty} \lambda^{n-1} K_n(x, y)$$

will certainly converge when $|\lambda| < N^{-1}$. The proof is equally valid when the interval of integration is infinite, provided that the integrals converge.

(E) Proof of Existence of Eigenvalue for Positive Definite Hermitian Kernel

If $\phi(x)$ and λ are respectively an eigenfunction and an eigenvalue associated with a kernel $K(x, y)$, then

$$\int |\phi(x)|^2 \, dx = \lambda \iint \bar{\phi}(x) K(x, y) \phi(y) \, dx \, dy$$

It will be convenient to write this in the short form $\bar{\phi}\phi = \lambda \bar{\phi} K \phi$. This short form will be used throughout the proof.

Let

$$I(\psi) = \bar{\psi} K \psi, \quad J(\psi) = \bar{\psi}\psi$$

Both $I(\psi)$ and $J(\psi)$ are real positive numbers depending on the arbitrary function $\psi(x)$.

The number $\Lambda(\psi) = J(\psi)/I(\psi)$ is thus also a positive real number depending on the function ψ. (The case of ψ identically zero everywhere is excluded.) Λ, I, J are termed functionals.

Construct a sequence of functions

$$\psi^{n+1}(x) = \int K(x, y) \psi^n(y) \, dy \quad \text{or} \quad \psi^{n+1} = K\psi^n$$

The exact form of ψ^0 is not important.

Let

$$J(\psi^n) = J^n, \quad I(\psi^n) = I^n, \quad \Lambda^n = J^n/I^n$$

If ψ^n is finite everywhere, so also is ψ^{n+1}. The same will be true for I^n, J^n and Λ^n.

Now

$$I^n = \bar{\psi}^n K \psi^n = \bar{\psi}^n \psi^{n+1}$$

and

$$I(\alpha\psi^n + \beta\psi^{n+1}) = \alpha^2 I^n + 2\alpha\beta J^{n+1} + \beta^2 I^{n+1} > 0$$

and

$$J(\alpha\psi^n + \beta\psi^{n+1}) = \alpha^2 J^n + 2\alpha\beta I^n + \beta^2 J^{n+1} > 0$$

if α and β are real.

It follows that

$$I^n I^{n+1} > (J^{n+1})^2$$

$$J_n J_{n+1} > (I^n)^2$$

and so

$$\Lambda^n = J^n/I^n > I^n/J^{n+1} > \frac{J^{n+1}}{I^{n+1}} = \Lambda^{n+1}$$

Thus Λ^n is a positive diminishing sequence and must tend to a limit λ_1 which is either positive or zero. It cannot be zero, because this would imply the existence of a non-identically zero ψ for which $J(\psi)$ is zero. λ_1 is thus positive and

$$\lim_{n\to\infty} J^n/I^n = \lambda_1$$

Also

$$\Lambda^n \geqslant \lambda_1 > 0$$

Now

$$\begin{aligned} 0 &\leqslant (\bar{\psi}^n - \lambda_1 \overline{K\psi}^n)(\psi^n - \lambda_1 K\psi^n) \\ &= \bar{\psi}^n\psi^n - 2\lambda_1\bar{\psi}^n K\psi^n + \lambda_1^2 \overline{K\psi}^n K\psi^n \\ &= \bar{\psi}^n\psi^n\left(1 + \frac{\lambda_1^2}{\Lambda^{n2}}\right) - 2\lambda_1\bar{\psi}^n K\psi^n \\ &\leqslant 2\{J^n - \lambda_1 I^n\} \end{aligned}$$

Now

$$\lim_{n\to\infty} J^n/I^n = \lambda_1$$

and so

$$\lim_{n\to\infty} \psi^n - \lambda_1 K\psi^n = 0$$

Suppose now that the functions ψ^n are normalized so that $J_n = 1$. The set ψ^n is bounded and the sequence

$$\psi^{n+1} = K\psi^n$$

is compact, and will have a limit function. That is there exists a limit function $\Phi_1(x)$ to the sequence $\psi^n(x)$. This is normalized so that $J(\Phi_1) = 1$. Passing to the limit, it follows that

$$\lambda_1 \int K(x, y)\Phi_1(y)\,dy = \Phi_1(y)$$

Thus, there exists an eigenvalue λ_1 with an associated eigenfunction ϕ_1, normalized so that $J(\Phi_1)$ is unity. It follows that

$$\lambda_1 \iint \bar{\Phi}_1(x)K(x, y)\Phi_1(y)\,dx\,dy = 1$$

(F) Convergence and Convergence in Mean

Consider the sequence defined by

$$C_n(x) = \sum_{p=1}^{n} c_p\Phi_p(x)$$

If

$$\lim_{n\to\infty} C_n(x) = C(x)$$

the sequence $C_n(x)$ is said to converge to $C(x)$. This is the normal process of convergence. If however this condition is not satisfied, but

$$\lim_{n\to\infty} \int |C(x) - C_n(x)|^2\,dx = 0$$

the sequence $C_n(x)$ is said to converge in mean to $C(x)$ over the domain of integration. Clearly a function which is convergent also converges in mean, but the converse is not always true. It may be said that $C_n(x)$ approximates $C(x)$ in mean.

Suppose that the $\Phi_r(x)$ form an infinite orthonormal set i.e.

$$\int \bar{\Phi}_r(x)\Phi_s(x)\,dx = \delta_{rs}$$

Consider

$$I_n = \int \left| f(x) - \sum_{r=1}^{n} f_r \Phi_r(x) \right|^2 dx > 0$$

where

$$f_r = \int f(x)\bar{\Phi}_r(x)\, dx$$

i.e. the f_r form the Fourier coefficients of the expansion in terms of the Φ_r.

It can easily be shown that

$$I_n = \int |f(x)|^2\, dx - \sum_{r=1}^{n} |f_r|^2$$

$$= \int |f(x)|^2\, dx - \sum_{r=1}^{\infty} |f_r|^2 + \sum_{r=n+1}^{\infty} |f_r|^2$$

Thus a necessary and sufficient condition that the sequence

$$\sum_{r=1}^{n} f_r \Phi_r(x)$$

converges in mean to $f(x)$ is that I_n tends to zero, that is

$$\int |f(x)|^2\, dx = \sum_{r=1}^{\infty} |f_r|^2 \quad \text{(Parseval's equation)}$$

and the orthonormal set $\Phi_r(x)$ is complete. If $f(x)$ and $g(x)$ are approximated in mean by

$$\sum_{r=1}^{\infty} f_r \Phi_r(x) \quad \text{and} \quad \sum_{r=1}^{\infty} g_r \Phi_r(x)$$

respectively, then $\alpha f + \beta g$ is approximated to in mean by

$$\sum_{r=1}^{\infty} (\alpha f_r + \beta g_r)\Phi_r(x)$$

Thus

$$\int |\alpha f + \beta g|^2\, dx = \sum_{r=1}^{\infty} |\alpha f_r + \beta g_r|^2$$

whence

$$\int \bar{f}(x) g(x)\, dx = \sum_{r=1}^{\infty} \bar{f}_r g_r$$

$$= \sum_{r=1}^{\infty} \bar{f}_r \int g(x)\bar{\Phi}_r(x)\, dx$$

(G) Complete Sets of Orthonormal Functions

Suppose that $\phi_r(x)$ are a set of functions which are orthonormal over a domain D, that is

$$\int \bar{\phi}_r(x)\phi_s(x)\,\mathrm{d}x = \delta_{rs}$$

(all integrals are over D).

If it is possible to expand an arbitrary function $f(x)$ defined over D in terms of these functions, let

$$f(x) = \sum_{s=1}^{\infty} f_s\phi_s(x)$$

The f_s may be evaluated as follows:

$$\int f(x)\bar{\phi}_r(x)\,\mathrm{d}x = \int \sum_{s=1}^{\infty} f_s\phi_s(x)\bar{\phi}_r(x)\,\mathrm{d}x = f_r$$

Consider now the series

$$\sum_{s=1}^{\infty}\left[\int f(x)\bar{\phi}_s(x)\,\mathrm{d}x\right]\phi_s(x)$$

Even if this series converges, it may not converge to $f(x)$. If however the series does converge almost everywhere to $f(x)$ as limit, then the set of orthonormal functions ϕ_r is complete. From Appendix F

$$\int |f(x)|^2\,\mathrm{d}x = \sum_{r=1}^{\infty} |f_r^2|$$

and

$$\int \bar{f}(x)g(x)\mathrm{d}x = \sum_{r=1}^{\infty} \bar{f}_r g_r$$

with an obvious notation. Thus if

$$\int |f(x)|^2\,\mathrm{d}x$$

is finite, the series

$$\sum_{r=1}^{\infty} f_r^2$$

will converge.

If the set of functions ϕ_r is incomplete it is clearly still true that

$$\sum_{r=1}^{\infty} f_r^2$$

will converge as all that has happened is that some of the terms have been left out of a convergent series. The meaning of the term complete is, in effect, that every single ϕ_r is needed for the expansion. If any are omitted, the series cannot represent an arbitrary function $f(x)$. This can be illustrated by considering the Fourier series. It is well known that if $f(x)$ has only a finite number of finite discontinuities in $-\pi < x < \pi$, then the series

$$\tfrac{1}{2}c_0 + \sum_{n=1}^{\infty} (c_n \cos nx + d_n \sin nx)$$

where

$$c_n = \frac{1}{\pi}\int_{-\pi}^{\pi} f(y)\cos ny\,\mathrm{d}y \qquad n \geqslant 0$$

$$d_n = \frac{1}{\pi}\int_{-\pi}^{\pi} f(y)\sin ny\,\mathrm{d}y \qquad n > 0$$

converges to

$$\tfrac{1}{2}\{f(x+0)+f(x-0)\} \qquad -\pi < x < \pi$$
$$\tfrac{1}{2}\{f(-\pi+0)+f(\pi-0)\} \qquad \text{at } x = \pm\pi$$

where $f(x+0)$ means $\lim_{\alpha\to 0} f(x+\alpha)$, $\alpha > 0$

The series thus converges to $f(x)$ except where $f(x)$ is discontinuous, that is almost everywhere. The set of orthonormal functions

$$\phi_0(x) = \frac{1}{\sqrt{(2\pi)}}, \quad \phi_{2s-1}(x) = \frac{1}{\sqrt{\pi}}\sin sx, \quad \phi_{2s}(x) = \frac{1}{\sqrt{\pi}}\cos sx \qquad s > 0$$

is thus complete over $-\pi < x < \pi$. If however one of the ϕ_r, say the constant term is omitted, then the series generated

$$\sum_{n=1}^{\infty} (c_n \cos nx + d_n \sin nx)$$

does not in general represent $f(x)$.

It follows immediately that, if

$$\int g(x)\bar{\phi}_r(x)\mathrm{d}x = 0 \quad \text{for all } r$$

then $g(x)$ vanishes almost everywhere. If however the set of functions ϕ_r is not complete, but could be completed by the addition of a set of functions ϕ_s^*, then

$$\int g(x)\bar{\phi}_r(x)\mathrm{d}x = 0$$

is satisfied by any $g(x)$ of the form $\sum_s g_s\,\phi_s^*$ where the g_s are arbitrary.

(H) Improper Integrals and Principal Values of Integrals

Consider the quantity

$$\int_a^b \frac{f(x)\mathrm{d}x}{x-c}$$

where $a < c < b$ and $f(x)$ is differentiable everywhere. With the usual definition of integral as a summation, this is undefined because of the singularity in the range of integration. Consider therefore the following:

$$I(\varepsilon) = \int_a^{c-\varepsilon} \frac{f(x)}{x-c}\mathrm{d}x + \int_{c-\varepsilon}^{b} \frac{f(x)}{x-c}\mathrm{d}x$$

where $\varepsilon > 0$, $c-\varepsilon > a$, $c+\varepsilon < b$. The integrals are perfectly well defined as there is no singularity anywhere. The singularity is being excised by an interval of length 2ε taken symmetrically about it.

$$I(\varepsilon) = \int_a^b \left\{\frac{f(x)-f(c)}{x-c}\right\}\mathrm{d}x - \int_{c-\varepsilon}^{c+\varepsilon} \frac{f(x)-f(c)}{x-c}\mathrm{d}x$$

$$+f(c)\left[\int_a^{c-\varepsilon} \frac{\mathrm{d}x}{x-c} + \int_{c+\varepsilon}^{b} \frac{\mathrm{d}x}{x-c}\right]$$

All these integrals are well defined. The apparent singularities at $x = c$ in the first two integrals do not matter as

$$\lim_{x\to c} \frac{f(x)-f(c)}{x-c} = f'(c)$$

which is finite. Also

$$\int_a^{c-\varepsilon} \frac{\mathrm{d}x}{x-c} + \int_{c+\varepsilon}^{b} \frac{\mathrm{d}x}{x-c} = \log\left(\frac{b-c}{c-a}\right)$$

Thus

$$I(\varepsilon) = \int_a^b \frac{f(x)-f(c)}{x-c}\mathrm{d}x + f(c)\log\left(\frac{b-c}{c-a}\right) - \int_{c-\varepsilon}^{c+\varepsilon} \frac{f(x)-f(c)}{x-c}\mathrm{d}x$$

The quantity

$$\lim_{\varepsilon\to 0} I(\varepsilon) = \int_a^b \frac{f(x)-f(c)}{x-c}\mathrm{d}x + f(c)\log\left(\frac{b-c}{c-a}\right)$$

is termed the principal value of

$$\int_a^b \frac{f(x)\mathrm{d}x}{x-c}$$

and is written

$$\int_a^{*b} \frac{f(x)\mathrm{d}x}{x-c}$$

(I) Generalized Functions

Consider the function which is defined as follows:

$$\begin{aligned} \delta_\varepsilon(x) &= K_\varepsilon \exp -\varepsilon^2/(\varepsilon^2 - x^2) \qquad & 0 \leqslant x^2 < \varepsilon^2 \\ &= 0 & \varepsilon^2 \leqslant x^2, \varepsilon > 0 \end{aligned}$$

and K_ε is defined by the relation

$$\int_{-\varepsilon}^{\varepsilon} \delta_\varepsilon(x)\mathrm{d}x = 1$$

$\delta_\varepsilon(x)$ has the following properties:

$$\begin{aligned} \int_a^b \delta_\varepsilon(x)\mathrm{d}x &= 0 \quad \text{if} \quad a, b \leqslant -\varepsilon \\ &= 1 \quad \text{if} \quad a \leqslant -\varepsilon, \varepsilon \leqslant b \\ &= 0 \quad \text{if} \quad \varepsilon \leqslant a, b \end{aligned}$$

Furthermore all of the derivatives of $\delta_\varepsilon(x)$ are continuous for all real x.

Suppose that $f(x)$ is an arbitrary function which obeys the relation

$$|f(x) - f(0)| \leqslant M|x| \quad \text{in} \quad |x| \leqslant \eta \geqslant \varepsilon \tag{A}$$

where M is a positive constant. Then

$$|f(x) - f(0)| \leqslant M\varepsilon \quad \text{in} \quad |x| \leqslant \varepsilon$$

Consider the quantity

$$\begin{aligned} I(\varepsilon) &= \int_{-\infty}^{\infty} f(x)\delta_\varepsilon(x)\mathrm{d}x - f(0) \\ &= \int_{-\varepsilon}^{\varepsilon} [f(x) - f(0)]\delta_\varepsilon(x)\mathrm{d}x \end{aligned}$$

Thus

$$|I(\varepsilon)| \leqslant M\varepsilon \int_{-\varepsilon}^{\varepsilon} \delta_\varepsilon(x)\mathrm{d}x = M\varepsilon$$

and

$$\lim_{\varepsilon \to 0} I(\varepsilon) = 0$$

That is

$$\lim_{\varepsilon \to 0} \int_{-\infty}^{\infty} f(x)\delta_\varepsilon(x)\mathrm{d}x = f(0)$$

$\delta_\varepsilon(x)$ vanishes outside $x = \varepsilon$, and as ε decreases the graphical representation of $\delta_\varepsilon(x)$ gets narrower and taller, with the area under the graph always remaining unity. Clearly ε cannot vanish as $\delta_\varepsilon(x)$ is not defined then.

If now the derivatives of $f(x)$ obey conditions similar to condition (A) above

$$\int_{-\infty}^{\infty} \delta_\varepsilon'(x)f(x)\mathrm{d}x = \left[\int_{-\infty}^{\infty} \delta_\varepsilon(x)f(x)\right] - \int_{-\infty}^{\infty} \delta_\varepsilon(x)f'(x)\mathrm{d}x$$

$$= -\int_{-\infty}^{\infty} \delta_\varepsilon(x)f'(x)\mathrm{d}x$$

Similarly

$$\int_{-\infty}^{\infty} \delta_\varepsilon^{(n)}(x)f(x)\mathrm{d}x = (-1)^n \int_{-\infty}^{\infty} \delta_\varepsilon(x)f^{(n)}(x)\mathrm{d}x$$

A quantity $\delta(x)$, the Dirac delta function, is now defined. It is not a function, but what is termed a generalized function or distribution. In a loose sense it is

$$\lim_{\varepsilon \to 0} \delta_\varepsilon(x)$$

$\delta(x)$ is defined as follows:

$$\delta(x) = 0 \qquad x \neq 0$$

and

$$\int \delta(x)f(x)\mathrm{d}x = f(0)$$

or alternatively

$$\int_a^b f(x)\delta(x)\mathrm{d}x = 0 \qquad a < b < 0$$

$$= 1 \qquad a < 0 < b$$

$$= 0 \qquad 0 < a < b$$

Similarly, the formal derivatives $\delta^{(n)}(x)$ are defined by

$$\delta^{(n)}(x) = 0 \qquad x \neq 0$$

$$\int_{-\infty}^{\infty} \delta^{(n)}(x)f(x)\mathrm{d}x = (-1)^n f^{(n)}(0)$$

The following properties hold:

$$\int_a^b f(x)\delta(x-\xi)\mathrm{d}x = 0 \qquad a<b<\xi$$

$$= f(\xi) \qquad a<\xi<b$$

$$= 0 \qquad \xi<a<b$$

$$\int_{-\infty}^{x'} \delta(x')\mathrm{d}x' = 0 \qquad x<0$$

$$= 1 \qquad x>0$$

This is the so-called Heaviside or unit function. It can be seen that

$$\int_0^\infty \mathrm{e}^{-px}\delta(x)\mathrm{d}x = 1$$

if the singularity lies just to the right of the origin. Also

$$\int_0^\infty \mathrm{e}^{-px}\delta^{(n)}(x)\mathrm{d}x = p^n$$

Thus $\delta(x)$ is the solution of the integral equation

$$\int_0^\infty \mathrm{e}^{-px}\phi(x)\mathrm{d}x = 1$$

Similarly

$$\int_{-\infty}^\infty \delta(x)\mathrm{e}^{i\omega x}\,\mathrm{d}x = 1$$

If $\phi_s(x)$ form a complete orthonormal set over some domain then formally

$$\delta(x-y) = \sum_{s=1}^\infty \phi_s(x)\phi_s(y)$$

as

$$\int \delta(x-y)f(y)\mathrm{d}y = \int \sum_{s=1}^\infty \phi_s(x)\phi_s(y)f(y)\mathrm{d}y$$

$$= \sum_{s=1}^\infty f_s\phi_s(x) \quad \text{where} \int f(y)\phi_s(y)\mathrm{d}y = f_s$$

$$= f(x)$$

In this sense $\delta(x-y)$ is the zeroth order iterated kernel.

References

1. CHAMBERS, LL. G. *An Introduction to the Mathematics of Electricity and Magnetism* (Chapman and Hall, London, 1973) p. 7.
2. WILLIAMS, I. P. *Matrices for Scientists* (Hutchinson, London, 1972) p. 57.
3. WILLIAMS, I. P. *Op. cit.* p. 62.
4. WILLIAMS, J. *Laplace Transforms* (George Allen and Unwin, London, 1973).
5. TITCHMARSH, E. C. *Introduction to the Theory of Fourier Integrals* (Oxford University Press, Oxford, 1937).
6. TRANTER, C. J. *Integral Transforms in Mathematical Physics* (Methuen, London, 1966).
7. MIKHLIN, S. G. *Variational Methods in Mathematical Physics* (Pergamon, Oxford, 1964) p. 230.

Further Reading

DELVES, L. M. and WALSH, J. (eds). *Numerical Solution of Integral Equations* (Oxford, 1974).

KANTOROVICH, L. V. and KRYLOV, V. I. *Approximate Methods of Higher Analysis* (Noordhoff, Groningen, 1958).

MIKHAILOV, L. G. *A New Class of Singular Integral Equations* (Wolters–Noordhoff, Groningen, 1970).

MIKHLIN, S. G. *Integral Equations* (Pergamon, Oxford, 1957).

MILLER, R. K. *Non Linear Volterra Integral Equations* (W. A. Benjamin, Menlo Park, 1971).

POGORZELSKI, W. *Integral Equations and their Applications* (Pergamon, Oxford, 1966).

RIESZ, F. and SZ-NAGY, B. *Functional Analysis* (Blackie, Glasgow, 1956).

TITCHMARSH, E. C. *Theory of Fourier Integrals* (University Press, Oxford, 1937).

TRICOMI, F. G. *Integral Equations* (Interscience, New York, 1957).

WIDOM, H. *Lectures on Integral Equations* (Van Nostrand–Reinhold, New York, 1969).

Index

MANSFIELD COLLEGE
WITHDRAWN
OXFORD